A DIAGNOSTIC PROCEDURE

- Start at box 1 and follow the appropriate arrows.
- To avoid a wrong or incomplete diagnosis, do not omit any steps.
- Remember that several different problems may be evident at once.
- For more information, see p.16

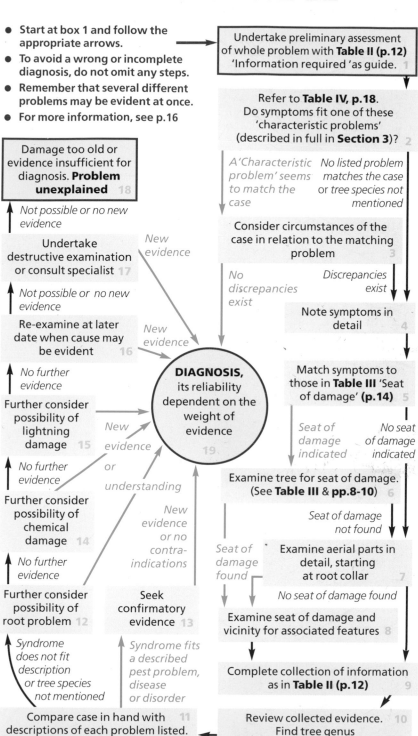

1 Undertake preliminary assessment of whole problem with **Table II (p.12)** 'Information required 'as guide.

2 Refer to **Table IV, p.18**. Do symptoms fit one of these 'characteristic problems' (described in full in **Section 3**)?

A 'Characteristic problem' seems to match the case — *No listed problem matches the case or tree species not mentioned*

3 Consider circumstances of the case in relation to the matching problem

No discrepancies exist — *Discrepancies exist*

4 Note symptoms in detail

5 Match symptoms to those in **Table III** 'Seat of damage' **(p.14)**

Seat of damage indicated — *No seat of damage indicated*

6 Examine tree for seat of damage. (See **Table III** & **pp.8-10**)

Seat of damage not found

7 Examine aerial parts in detail, starting at root collar

Seat of damage found — *No seat of damage found*

8 Examine seat of damage and vicinity for associated features

9 Complete collection of information as in **Table II (p.12)**

10 Review collected evidence. Find tree genus and seat of damage in Host Agent directory (**Section 2**)

11 Compare case in hand with descriptions of each problem listed. If none matches, turn to **'Many Tree Genera' (p.24)** and repeat process.

Syndrome does not fit description or tree species not mentioned — *Syndrome fits a described pest problem, disease or disorder*

12 Further consider possibility of root problem

13 Seek confirmatory evidence

14 Further consider possibility of chemical damage

No further evidence

15 Further consider possibility of lightning damage

No further evidence

16 Re-examine at later date when cause may be evident

Not possible or no new evidence

17 Undertake destructive examination or consult specialist

Not possible or no new evidence

18 Damage too old or evidence insufficient for diagnosis. **Problem unexplained**

19 DIAGNOSIS, its reliability dependent on the weight of evidence

New evidence

New evidence

New evidence or understanding

New evidence or no contra-indications

DIAGNOSIS
OF ILL-HEALTH
IN TREES

This book is dedicated to the memory of
C.W.T ('Bill') YOUNG (1920–1980)
who taught R.G.S. to see
what he was looking at.

DIAGNOSIS OF ILL-HEALTH IN TREES

by R.G. Strouts
and T.G. Winter

Published by TSO (The Stationery Office) and available from:

Online www.tso.co.uk/bookshop

Mail, Telephone, Fax & E-mail
TSO, PO Box 29, Norwich, NR3 1GN
Telephone orders/General enquiries: 0870 600 5522 Fax orders: 0870 600 5533
E-mail: book.orders@tso.co.uk Textphone 0870 240 3701

TSO Shops
123 Kingsway, London, WC2B 6PQ 020 7242 6393 Fax 020 7242 6394
68-69 Bull Street, Birmingham B4 6AD 0121 236 9696 Fax 0121 236 9699
9-21 Princess Street, Manchester M60 8AS 0161 834 7201 Fax 0161 833 0634
16 Arthur Street, Belfast BT1 4GD 028 9023 8451 Fax 028 9023 5401
18-19 High Street, Cardiff CF10 1PT 029 2039 5548 Fax 029 2038 4347
71 Lothian Road, Edinburgh EH3 9AZ 0870 606 5566 Fax 0870 606 5588

TSO Accredited Agents (see Yellow Pages)

and through good booksellers

Following the reorganisation of the government in May 2002, the responsibilities of the former Department of the Environment, Transport and the Regions (DETR) and latterly Department for Transport, Local Government and the Regions (DTLR) in this area were transferred to the Office of the Deputy Prime Minster.

First Published 1994
Second impression 1998

Second Edition 2000
Second impression 2004

ISBN 0 11 753545 1

CONTENTS

PREFACE

We offer this book to arboriculturists and others who do not have specialist pathological or entomological knowledge, in the hope that it will satisfy their need not only to reach well-founded diagnoses but also to take appropriate action thereafter.

At the same time, we have also sought to provide a uniquely comprehensive yet practical summary of the tree problems that are likely to be encountered by the arboriculturist. Understanding that a mere catalogue of problems is of limited practical value and can be bewildering, we have distinguished those which require or which will respond to control measures from those which can be safely ignored or cannot be controlled.

In addition to the well-known and well-researched pest, disease and abiotic problems, we describe a number which are unfamiliar or previously undescribed including some for which explanations are still lacking. We hope that readers will bring further examples of these to our notice so that their distribution and importance can be better judged.

We also hope that this book will serve as a useful and interesting introduction to the subject of tree pathology in its widest sense.

We apologize for any omissions or errors that have escaped us and ask that these, especially errors of fact, along with suggestions for improvement, are brought to our notice.

Robert Strouts, *Pathologist*   **Tim Winter,** *Entomologist*
Alice Holt Lodge, May 1994

PREFACE to 2nd Edition

This second edition retains its original structure but we are pleased to incorporate a number of minor, useful modifications to facilitate its use as suggested by users of the first edition, notably the highlighting of references in the text to Figures, and improvements to the Index.

We are disappointed that since 1994 no new information has become available on the several unexplained diseases that we describe. Since then, however, several conditions have become prominent enough to warrant full entries in Section 3 (Phytophthora disease of alder, Phytophthora blight of holly, Oak dieback) and a few new pests and diseases which have begun to trouble amenity trees are mentioned briefly in Section 2 (Cylindrocladium blight of box, Anthracnose of Flowering dogwood, Pittosporum sucker).

A major new addition to Section 3 is a guide to the identification of the principal wood-rotting fungi on standing trees. A few other subjects not dealt with earlier have also been added, notably Waterlogging and Exotic pests and diseases.

Again, we apologize for any errors new or old that have escaped us and ask that these be brought to our notice. And we hope that this enlarged version will still fit into the arboriculturist's pocket.

Robert Strouts, *Pathologist*   **Tim Winter,** *Entomologist*
Liss, Hampshire   **Haslemere, Surrey**
April 2000

ACKNOWLEDGEMENTS

FIRST EDITION

This book was written as part of a contract placed with the Forestry Commission by the Department of the Environment's Directorate of Rural Affairs. It is a distillation of information from many sources including the case histories of investigations into tree problems built up over 30 years at the Forestry Commission's research stations. We acknowledge with pleasure the huge contribution our many colleagues have made to this store of knowledge, accumulated during and before our long terms as advisory officers with the Commission's Pathology and Entomology Branches, in particular the Pathology Advisory Service (South)'s succession of mycologists: Mrs C. G. Gulliver, Mrs S. M. Dennis and Mrs T. C. Reffold.

All books of this nature are built on the earlier work of many researchers and authors. The principal textbooks we have drawn on are listed in the bibliography. Without those and the many other books and published papers we have consulted, this work could not have been accomplished.

Mention should also be made of the innumerable arboriculturists, foresters and gardeners whose inquiries over the years have provided the purpose and wherewithal for our work.

We thank all those who made so many helpful and tactful suggestions for improvements in our early drafts, in particular our colleagues, Dr J. N. Gibbs, Dr H. Evans and Mr D. Patch; Mrs C. Davis of the Department of the Environment; and the arboriculturists Mr D. Thorman and Dr P. G. Biddle. For useful comments on certain subjects we thank Dr J. I. Cooper (viruses), Dr A. T. K. Corke (Spruce bud blight), Dr D. Lonsdale (decay and safety), Dr A. J. Moffat (lime induced chlorosis), Mr C. Carter (aphids) and Dr L. Sutherland and Mr B. J. W. Greig (Dutch elm disease).

We also acknowledge the invaluable assistance given to us by our typists, clerks, and photographic staff, especially our present Principal Photographer, Mr G. Gate, and Mr D. R. Rose, who not only undertook most of the pathology advisory work during the preparation of this book, but also so often and so patiently came to the aid of a computer-ignorant author.

Most photographs were taken either by the authors or by past and present members of our Photography, Pathology and Entomology Branches, to all of whom we are grateful. The majority are in the Forestry Commission's Research Division collection. We also thank the following for the use of their photographs, as stated: Dr J. I. Cooper, Institute of Virology and Environmental Microbiology, NERC, Oxford (**Fig.51**); Dr A. Soutreneon, CEMAGREF, Division Protection Phytosanitaire de la Forêt, Grenoble, France (**Fig.24**); the Ministry of Agriculture, Fisheries and Food (**Fig.138**); Dr D. V. Alford of Cambridge (**Fig.186**); Mr A. W. Brand of Stratford-upon-Avon (**Fig.324**); Mr. A. R.Outen of Hitchin (**Fig.317**) (the last two supplied by the British Mycological Society).

Finally, we thank Mr J. L. Angell and his staff at the Department of the Environment and staff at HMSO for allowing us so much say in the book's design; in particular we thank Mrs J. Hannaford for her skill and patience in translating our ideas into an attractive reality.

ACKNOWLEDGEMENTS

SECOND EDITION

We thank all those users of the first edition who have been kind enough to compliment us on its value in their work and those who have contributed to the second edition by pointing out mistakes and suggesting improvements.

We are grateful for the technical information supplied by Dr. C. Prior (Cylindrocladium blight of box and Anthracnose of Flowering dogwood), Mr A. Halstead (Pittosporum psyllid), and Dr. J.N. Gibbs (Phytophthora disease of alder) and acknowledge the continuing invaluable work of the Forestry Commission's Pathology and Entomology Branches both North and South from where so much of the information in this book comes.

Our warm thanks are also due to Mrs. M. Peacock for so patiently transferring our fragmentary manuscripts and typescripts to computer disk; to the printers Graphics Matter Ltd., for incorporating the many additions and amendments so skilfully and thoughtfully into the original book; to Dr. J.N. Gibbs for inviting us, in our retirement, to undertake the revision and for facilitating its progress to completion; and to Eileen, the wife of the first author, for allowing him, without complaint, to neglect his domestic duties so often and for her help in proof reading in a subject not very close to her heart. Most of the new photographs have been provided by staff of the Forestry Commission Research Agency or by the first author. We also thank the following: Royal Horticultural Society (Fig.24 and Fig.C3), Tim Sandall (Fig.99) and Dr. C. Prior (Fig.C8).

INTRODUCTION

USING THE BOOK

The book is designed to be used as an integrated whole.

It is in three main sections: (1) Diagnostic procedure, (2) a directory arranged by tree genus to the problems described, and (3) detailed illustrated descriptions, with commentaries, of the principal pests, diseases and disorders included. There are also other sections on control measures and on decay and safety.

Initially, the user is urged to become familiar with the content of Section 1. In this, we have sought to systematize the diagnostic procedure which the experienced investigator uses (albeit probably unconsciously and haphazardly).

The Diagnostic Procedure, as outlined in the flow chart on p.17, should be followed, step by step, during each investigation. By systematic study of symptoms and circumstances, more and more possible causes can be eliminated until the problem has been characterized adequately and the seat of damage ascertained. The Host-Agent Directory is then consulted and each problem on the short list matched to the case in hand. If a good match is not found, further steps to resolve the question are suggested.

SCOPE OF THE BOOK

(a) Geographical Area Covered

All the problems described in this book occur somewhere in Great Britain. If the pest or disease is known to be confined to a particular region of the country this is stated but most are widespread and, given suitable conditions, could be encountered wherever the host tree is grown. Although most of the problems also occur in Ireland and in Continental Europe (and many of them in North America and other temperate regions of the world), some do or may not, so particular care should be taken if the book is used outside England, Scotland and Wales.

(b) The trees included

This book deals with the problems affecting trees which have been planted out for the purpose of ornament, screening, shading or hedging in Great Britain.

The tree genera ('host' genera) dealt with in this book are nearly all of those included by A. F. Mitchell in his Field Guide to the Trees of Britain and Northern Europe (published by Collins 1974) in which he defines a tree as a woody perennial '. . .that commonly achieves a height of 6 metres on a single stem. . .plus hazel' (*Corylus avellana*). Information on the pathology of many of these genera is scanty or entirely lacking but it is precisely for this reason that they have been included: we hope that not only will this limited information prove useful but also that any problems encountered by readers on these less familiar trees will be brought to our attention and so help add to our knowledge of their pathology.

(c) The pests, diseases and disorders included

Ill health due to every kind of agent likely to affect amenity trees in this country is discussed. Problems peculiar to nurseries and forestry are not included.

With the exception of damage due to mammals and biting insects, which we have included only if it is commonly misidentified (Poplar leaf beetle damage and squirrel damage to sycamore branches, for example), the disorders, diseases and pest problems described in detail include the vast majority of those likely to be encountered. A few problems about which little is known or which have not yet been fully characterized (e.g. Diaporthe canker of Laurus; Ash dieback) and the more obvious mechanical damage caused by mammals and biting insects (bark stripping by horses and caterpillar damage to leaves, for instance) are given only passing mention in the Host-Agent directory, or are dealt with in general terms in Section 1 or under 'Mechanical Damage' in Section 3.

Our decision whether to include or exclude any particular pest, disease or disorder has been based largely on the frequency with which it has been brought to the attention of the Forestry Commission's advisory services, in relation to the commonness of the tree species involved. Some apparently uncommon problems with features of special interest are also included.

Some tree genera have figured very little or not at all in the enquiries we have received over the years. For pests and diseases more or less confined to forestry plantations, the reader is referred to Gregory and Redfern (1998) and Bevan (1987). Problems peculiar to nurseries are not included.

CONVENTIONS

For the sake of brevity and to avoid ambiguity, the Latin names of trees, fungi, insects and other organisms have been favoured over the vernacular names. The latter are given in the Index with their Latin equivalents.

Also for the sake of brevity, though perhaps at a slight risk of ambiguity, authors of scientific names have not been quoted. The names used are those current at the time of writing; their authors will be found in, for example, Phillips and Burdekin (1992), Mitchell (1974) and Winter (1983).

To facilitate the reading of the text and to enable italics to be used unequivocally for emphasis or subheadings, the conventional use of italics for Latin names of trees and pathogens has been abandoned in Section 3, the Descriptions of Pests, Diseases and Disorders.

PESTICIDES

It should be noted that the mention in this book of any pesticide (fungicide, bactericide, insecticide, acaracide, herbicide) does not imply that that material or product is available, nor that, if available, it is currently licensed for the purpose in question. The reader is referred to Section 4, Item C. Pesticide Regulations, on pages 288-290.'

SECTION 1

AN INTRODUCTION TO THE DIAGNOSIS OF ILL-HEALTH IN TREES

AN INTRODUCTION TO THE DIAGNOSIS OF ILL-HEALTH IN TREES

WHAT CONSTITUTES ILL-HEALTH IN A TREE?

As in a human, ill-health in a tree implies that the organism as a whole is in some way malfunctioning. For this book, however, ill-health is taken to include any supposed deviation from the normal however much or little of the plant is affected.

WHY DIAGNOSE THE PROBLEM?

An accurate diagnosis provides the necessary basis for the implementation of appropriate control measures and compliance with current plant health legislation; it also enables those problems which can and need to be controlled to be distinguished from those which cannot and need not. Until the nature and cause of a problem is clear, its importance cannot be assessed: the death of a small twig may be an insignificant event, or it may be the first sign of a fatal disease.

Provided that good records are kept, diagnoses also provide those who research and advise on tree pathology with much useful information on the distribution, incidence and effects of many pests, diseases and disorders, which would otherwise be lost.

In addition to these practical reasons, there is considerable intellectual satisfaction to be had from pursuing a diagnosis for its own sake, as in the unravelling of any puzzle, while the recognition and understanding of tree pests, diseases and disorders can only enhance one's appreciation and enjoyment of the living world.

PREREQUISITES FOR A DIAGNOSIS

A certain **basic knowledge of tree pathology** is of course indispensable to the would-be diagnostician. It is necessary to be aware of the various things which damage trees, to understand how they inflict damage, and to be able to deduce from symptoms which part of the tree has suffered the primary damage. These matters are discussed briefly below. Equally necessary is a broad **understanding of tree biology** and the **ability to identify trees**. If the reader is deficient in either respect, the authors recommend the study of the introductory chapters of a book such as A. F. Mitchell's *Collins Field Guide to the Trees of Britain and Northern Europe*. One can become familiar with the terms and processes described by examining a few conifers and broadleaved trees and observing their changing state through the seasons.

For those who wish to study tree pathology in more depth, the introductory and general chapters in T. R. Peace's *The Pathology of Trees and Shrubs* are recommended (see bibliography; now available only in libraries), and the American book, *Diseases of Shade Trees* by T. A. Tattar, published by the Academic Press, Inc., revised 1989.

TOOLS AND EQUIPMENT

Rarely can a problem be adequately investigated without the need to cut off twigs, cut into bark or to examine roots. Much can be done with no more than a stout, sharp knife, a strongly-made trowel and moderately-powered hand lens, but often, especially for larger trees and when detailed examinations are necessary, other tools and equipment are required. The following kit has stood the test of time for the authors:

Hand lens (8× or 10× magnification is sufficient).

Binoculars (7× or 8× magnification is adequate; 50mm objective for good performance in poor light; as close focusing as possible).

Pruning knife.

Secateurs.

Wood chisel and mallet.

High pruners.

Pruning saw, short handled.

Bow saw, 18″ blade.

Trowel (strong blade-to-handle fixing, narrow).

Combination axe/mattock, such as the VB Groundbreaker.

Extending ladder (to 12ft sufficient, convertible to a stable step-ladder).

Note book, pen, pencil.

Polythene bags and closures.

Containers for insects (rigid, with secure lids).

Labels (for specimens and specimen bags).

Felt-tip pen, spirit-base.

Wound paint, fungicidal (see Sections 4 and 5).

Protective clothing (helmet, goggles, gloves).

Some additional useful items:

Hatchet; entrenching shovel; lightweight pick-axe; tape measure; polythene food-storage boxes; camera with colour film; compass (for checking orientation of damage); Pressler increment borer (for dating bark injury).

And if decay detection is undertaken, the appropriate specialised equipment (see Section 5).

For one's own welfare and for insurance and legal reasons, it is essential that one observes current safety and health regulations whilst working. Nowhere is this more important than when climbing ladders or trees or working with cutting tools. Fortunately (and perhaps surprisingly) it is rarely necessary to climb in order to investigate most problems adequately. When it must be done, it may be necessary or wise to employ a trained tree climber.

TREE-DAMAGING AGENTS

The causes of damage and abnormalities can conveniently be divided into two broad categories: those which are living organisms (biotic agents) and those which are non-living (abiotic). These may be usefully further divided up as in Table I on p.11.

HOW THESE AGENTS DAMAGE TREES

Mammals strip bark off with teeth, claws and horns; they gnaw bark above the ground and on roots; they break or bite through twigs and shoots and eat shoots, leaves and fruits; they trample and compact the ground in which tree roots are growing. *Man* may inflict direct chemical or mechanical damage or harm trees by cultural malpractice.

Birds break shoots by alighting on them; peck out buds; drill holes and make peck marks in bark.

Biting insects consume foliage, roots and bark; gnaw bark above and below ground; tunnel under bark and in fruit, seeds, leaves, buds, shoots, roots and wood; graze the superficial layers from leaves. A few cause swellings and other growth abnormalities. Many are large enough to be seen though many are elusive.

Even though mammals, birds and biting insects are living, the damage they cause is entirely mechanical and is discussed further under that heading below.

Sucking insects and mites obtain their nourishment as a liquid extracted from phloem and other soft tissues. The tissues may become discoloured, abnormal outgrowths and other growth distortions may result and the tree may be poisoned or weakened in the process. Many of these creatures are elusive and others invisible or barely visible to the naked eye.

Nematodes (eelworms – microscopic worms) mostly feed by sucking out cell contents. The result is usually stunted growth and growth abnormality. Exceptionally they live inside and block vascular xylem tissues and bring about wilt disease. They have very rarely been implicated in the ill health of amenity trees in this country.

Many *animals*, large and small, carry pathogens from tree to tree.

Fungi. Most plant pathogenic species grow through or over the tissues they feed on as microscopic cellular threads (hyphae); a few are carried passively in the tree's water conducting vessels. Invaded living tissues are usually killed sooner or later, while prolonged activity of certain fungi in the wood of the tree results in its decay. Some fungi live within the water conducting tissues of the wood, disrupting the tree's water supply and sometimes also poisoning the tree; other fungi cause growth abnormalities. Many species produce visible fruit bodies or other structures or their massed hyphae (mycelium) can be seen in or on the plant.

Bacteria are microscopic, single-celled organisms which, by feeding, multiplying and moving (actively or passively) in the tree's vascular system kill plant tissues or induce growth abnormalities. Some species sometimes become visible as mucilaginous drops oozing from infected tissues.

The so-called *rickettsia-like organisms* of older plant pathological literature are now known as *xylem-limited bacteria* (XLB's), or *fastidious xylem-inhabiting bacteria* (FXIB's). In recent years they have been found to be responsible for some tree diseases characterized by interveinal and marginal leaf necrosis, but not yet in this country. The rickettsia bacteria differ from these and are known animal but not plant pathogens.

Spiroplasmas, Mycoplasmas and Phytoplasmas are, like bacteria, single-celled organisms but differ from bacteria in having non-rigid cell walls, and in their biochemistry. *Spiroplasmas* are known to cause a few plant diseases (including one of citrus trees) but *mycoplasmas* are known only as animal pathogens. Organisms resembling mycoplasmas (until the 1990s known as *mycoplasma-like organisms* or MLOs, now named *phytoplasmas*) are, however, known to be the cause of some 'yellows' diseases. The certain or likely causes of several plant diseases previously attributed to viruses have, since about 1970, been found to be XLB's or MLO's. Both kinds of organism are likely to assume greater prominence in plant pathology as research continues.

Viruses are subcellular particles, much smaller than most bacteria. They replicate only in the living cell and in so doing interfere with its normal function, giving rise to diseases which are variously characterised by growth-rate reduction, growth abnormalities, discoloration of soft plant tissues and graft failure.

Viroids are non-particulate assemblages of ribonucleic acid (RNA), first shown to cause disease in plants in 1971. To date they are suspected of causing disease in some tropical trees.

Higher Plants. The only parasitic higher plant to cause visible damage to trees in this country is the mistletoe, *Viscum album*, a semi-parasite, and the damage this causes is usually negligible. It derives water and minerals from the host tree through its 'roots' which ramify through the bark down to the wood, but it has green, photosynthesizing leaves of its own. Sometimes, however, instead of the familiar leafy balls (Fig. 94), only a few leaves sprout from greatly swollen branches (p. 104, Fig. 98), probably an indication that the plant is depending heavily on the host tree for its carbohydrates as well. Ivy, *Hedera helix*, is often mistakenly thought to parasitize trees, but its rootlets merely enable it to climb up the tree (or wall). Only in trees naturally thin-crowned or rendered so by disease or disorder will ivy pose a threat to its host's health, by smothering the tree's foliage with its own in time.

Drought. Water shortage reduces the uptake of minerals and impairs photosynthesis. Leaves may be shed to reduce water loss and bark and wood shrinkage may cause longitudinal cracking.

Flooding and waterlogging reduce the oxygen available to roots. These function less and less efficiently and eventually die, causing consequent crown deterioration. Salt water is in addition toxic to most trees.

Spring and autumn frosts and winter cold. Freezing dehydrates, ruptures and so kills tissues which are inherently or temporarily in a cold-sensitive physiological state.

High temperatures commonly exacerbate drought damage. In extreme cases, such as in fire damage, living cells may be killed by heat alone.

Winds in summer contribute to drought conditions. In winter too, drying winds may remove water from evergreen leaves faster than it can be replaced from cold soil and so induce drought symptoms. Winds also break stems, branches and roots, and foliage may be torn or thrashed and damaged against neighbouring objects. Many pathogenic organisms are wind-disseminated; and wind can carry herbicidal sprays to non-target plants or deposit damaging salt spray onto plants.

Snow and ice weigh down and break or permanently bend branches and stems.

Lightning. Trees may be shattered or fissured when the heat generated in a discharge turns the water in wood and bark instantly into steam. Trees which die without such physical injury have presumably been subjected to less powerful discharges.

Harmful chemicals may interfere with a tree's metabolic processes, killing tissues or stunting or distorting growth. Some (e.g. oils) may form a layer impervious to air or water over root or leaf surfaces. The roots of Juglans, and perhaps of a few other species, produce a toxin inhibitory to the survival of the seedlings of certain (but not all) other plants which germinate in their vicinity. This effect (termed allelopathy) cannot, however, be disentangled in the field from the inhibitory effects of competition for light, water and minerals.

Agents of mechanical damage. Impacts from vehicles, implements, shotgun pellets, hail and other flying objects, wind, and the weight of snow and ice may break, abrade, cut or tear any part of the tree. Most of the damage caused by mammals, birds and biting insects also falls into this category. The wounds may allow infection by fungi or bacteria. Heavy machinery and large animals can compact soil and so reduce the oxygen available to roots that their function is impaired. The crown consequently deteriorates. The addition of a thick or impervious layer of material over existing roots can have a similar effect. Neglected tree ties and other constrictions become ever tighter as the tree grows, eventually so restricting the production of new phloem and xylem that photosynthesized materials cannot reach the roots and water cannot reach the foliage.

Dust deposits may be phytotoxic (see 'Harmful chemicals' above) or may impede photosynthesis by reducing the light and air reaching the leaves.

Street Lighting. Plants respond to light mainly in the red or far red regions of the spectrum. Much street lighting consists mainly of light of shorter wavelengths or is of an intensity too low to affect plant growth, but high pressure sodium lamps do emit much high intensity red light and can, if very close to trees, prevent flowering or induce growth late into the autumn. Such unhardened shoots may then be killed by autumn frosts (see Tattar 1978). Genera listed by Tattar as particularly susceptible include Acer, Betula, Catalpa, Platanus and Zelkova.

'Planting losses'. Whatever the mistreatment or misfortune which leads to planting losses where pests and diseases are not to blame, the fundamental cause will have been chemical or mechanical damage, dryness or an excess of water, overheating or freezing, or a combination of these factors, any of which may have occurred at any time from lifting until after planting. *Over-large planting stock*, unless especially prepared, planted and tended as such, often grows feebly for years or slowly dies. The problem is particularly likely to arise in conifers and in trees transported long distances from abroad. See also *Weed competition* below.

Nutrient deficiencies. A shortage of or imbalance in the elemental nutrients derived from the soil will inhibit various of the plant's metabolic and physiological processes, such as the synthesis of chlorophyll and enzymes and the uptake of water, and thus affect all aspects of the tree's health and growth. Symptoms range from leaf discoloration to stunted growth.

Weed competition, especially a grass sward, deprives the newly planted tree of moisture and therefore also of plant nutrients. Growth may be poor for many years.

Graft failure/incompatibility. Incomplete fusion of scion with rootstock sometimes leads to the snap of the tree stem at the union or the slow decline and death of the scion, even after a considerable number of years of healthy growth. Often the still healthy root stock throws up healthy shoots. Such imperfect unions are most likely where rootstock and scion are of different genera or as the result of virus infections.

Replant diseases, soil sickness. Sometimes, despite their good quality and the care they have received since leaving the nursery, newly planted trees (and shrubs) fail to thrive and may even die quite soon with no clear-cut sign of attack by pest or disease. Especially, but not only, if the species involved is *Cydonia, Malus, Prunus* or *Pyrus* and is a replacement for one of these species on the same site, a 'replant disease' can be suspected. Little, if any, new root growth will be evident on affected trees. The causes have not been fully elucidated but are not always the same; they often involve various root-killing fungi and eel-worms. Trees usually recover if transplanted to a new site or if the suspect soil is changed. For more information consult Buckzacki (1998), ADAS or (for RHS members) the Royal Horticultural Society (addresses appear on p. 306).

Combinations of factors (Complex diseases; declines). A sudden and severe stress from one of the damaging agents listed above (e.g. complete defoliation by insects) or a chronic or repeated stress (such as a nutrient deficiency or perennial waterlogging) may, by interfering with metabolic or physiological processes, severely impair the tree's ability to cope with any further stresses to which it might be subjected. If adverse conditions or pest or disease attacks continue the problem is compounded, the tree becomes progressively weaker and less able to resist even weak pathogens and eventually dies from this interaction of several different factors.

THE INGREDIENTS FOR A SUCCESSFUL DIAGNOSIS

- Before immersing yourself in the examination of tree and site, make time to take in the scene, and during the investigation pause now and then to look around quietly and to reflect on what you have found so far.
- Make your observations deliberately and systematically; do not rush – it is all too easy to overlook an important detail. Check against Table II, column 1 (pp.12–13) to ensure that you have considered all aspects of the problem.
- Avoid the temptation to jump to conclusions before the investigation is complete. You are bound to formulate hypotheses as you accumulate information but the last observation you make can change your mind entirely.
- Expect the explanation for the damage to be straightforward – it usually is.
- If your diagnosis does not quite fit the facts, question it.

UNDERSTANDING SYMPTOMS

Although the visible symptoms are the manifestation of the disorder, they are rarely distinctive enough to enable an accurate diagnosis to be made, as particular symptoms can have many different causes. For example, leaves may fall as the result of fungal or bacterial infection, damage from biting or sucking insects, drought, herbicides, wind or root disease.

On their own, then, symptoms do not usually provide the explanation for the tree's ill health. Their study is, however, the first step in the diagnostic process, as they often indicate where the seat of the problem lies, that is, where the damage which has elicited the symptoms has occurred. Once that place has been located, the nature of the primary damage (death or girdling of roots, for example) can be discovered and the site of the damage can be examined for the causal agent. The likely seats of damage, as indicated by the gross symptoms, are listed in Table III on p.14.

EXAMINING THE SEAT OF DAMAGE
(SEE ALSO P.293, INSPECTING TREES FOR SIGNS OF DECAY)

Once the probable seat of damage has been deduced from symptoms, that part of the tree must be examined for dead, dying, missing, discoloured or deformed tissues. If these are found, (confirming that this is indeed the probable seat of damage), their character and distribution should be noted, and note made also of any associated features (see Table II, 3 The Problem, (p.13)). If you are unsure what healthy tissue should look like, examine an evidently healthy part of the same tree or a healthy tree of the same species.

Roots and Root Collar

Roots are the most difficult part of the tree to examine and so are often ignored or given cursory attention. However, a systematic examination with a minimum of digging often proves fruitful and, in all cases where symptoms admit of a root problem, should be undertaken.

(i)   Start by examining the stem at soil level closely for fungal fruit bodies. These may be small and inconspicuous or very close to the ground and concealed by vegetation; the presence of a known pathogenic species will add much weight to or may suffice to confirm a tentative diagnosis of root disease.

(ii) Especially if no fruit bodies are found, examine the stem base again for dead bark, starting at any exudate or between root buttresses (where dead bark extending up from roots often first appears). It is very often impossible to distinguish dead from live bark from a superficial examination; therefore, at regular intervals around the tree, prise up or cut out with a chisel or knife small pieces of outer bark to check the condition of the inner bark adjacent to the wood (on thin-barked trees, tiny cuts may suffice).

(iii) If dead bark is found, cut further to check for fungal mycelium in and under the bark or in the wood; check the wood for decay; ascertain the extent and configuration of the dead bark.

If no dead bark is found:

(iv) Examine the stem base again, this time a few inches below ground level; include the major visible roots in the examination.

If still no dead bark is found:

(v)   Attempt to examine deeper roots. Select four points equally spaced around the stem and a few feet from it. Dig a narrow hole as deep as practicable at each place, working between any sizeable roots and checking the condition of all roots encountered. Note also features of the soil's texture, odour or colour which might indicate an inhospitable rooting environment, such as a raised soil level, chemical contamination or waterlogging.

If no explanation for the damage has been found by now but a root problem still seems to be worth pursuing:

(vi) Further digging may be revealing. Start against the stem, between root buttresses. A stout trowel is usually the most satisfactory implement for this purpose. Then repeat step (v) but a little further away from the stem.

How much more digging can be done depends on the soil type and condition, the number and type of roots and on the time and patience available. At best only a tiny proportion of the root system of a large tree can be examined in this way so failure to find dead, dying or decayed roots does not rule out the possibility that they are present but inaccessible.

Bark and Cambium of Stem, Branches and Twigs.

(i)   Examine the bark proximal to (i.e. below) the wilted or dead parts: it may be missing, roughened, sunken, swollen or cankered. If so, note any exudations or other associated features. Dead bark that has lifted away from the wood beneath will yield and sound hollow if tapped with a mallet.
If the bark appears to be normal:

(ii) Starting just proximal to the affected part, gently prise up or cut out small pieces of outer bark to check the condition of the inner bark. Check the extent and configuration of any dead bark and of any stain in the underlying cambium; check for decay and stain in the wood; note any associated features.

If only healthy bark is found:

(iii) Repeat the process at short intervals up and down the member, starting in the vicinity of the affected part. If still only healthy bark is found, proceed as in the next paragraph.

Water-conducting Wood

Beginning at the distal end of the affected branch and working back towards the tree stem, cut into the wood at intervals of 6 or 12 inches and examine it for stains of the kind described under Verticillium wilt or Dutch elm disease (p.261 and p.116). The stain which is characteristic of such vascular wilt diseases may be present only at the very base of wilted branches, or even only in the stem. If permissible, such examinations are best carried out by cutting the branch back piece by piece with secateurs or saw. The cut ends often need trimming with a clean, sharp knife before examination. Check the cambial area by paring the bark away.

Foliage

Examine shoots, buds (cut through some), stipules, petioles and both sides of leaves for discoloured or necrotic areas; note their character and distribution and any associated features.

SHORT CUTS TO A DIAGNOSIS

Early on in the Diagnostic Flow Chart on p.17, the investigator's attention is drawn to a list of problems characteristic of certain tree species (Table IV, p.18); many steps in the diagnostic process can sometimes be circumvented by comparing the symptoms exhibited with the symptoms described in Section 3 for these characteristic problems. Such short cuts must be used circumspectly as it is easy to jump to wrong conclusions if due consideration is not given to all the circumstances of the case.

IF THE INVESTIGATION FAILS

Do not be surprised if, after all your efforts, you cannot explain the damage. Failure may be because insufficient information is available (often an investigation begins so long after the initial damage that the cause and evidence for it has vanished); or it may be due to the investigator's inexperience or lack of knowledge; or the disorder may not be described in the books to hand; or, rarely, the problem may be a new or undescribed one: much remains to be discovered in the field of amenity tree pathology.

In some cases, for practical reasons, it may not be possible to pursue an investigation to a conclusion: roots may be inaccessible or the owner may not sanction the amount of damage to the tree or garden that a thorough examination might entail. Whatever the reason for failure, some useful information will have been acquired. It is not helpful, however, to speculate on the cause and it is positively unhelpful to present guess-work as certainty. Often, failure to find the precise cause matters little as the investigation may well have provided the information necessary to answer the main concerns, which are usually fourfold: 'Will it spread to other trees?'; 'Will the tree die?'; 'Is the tree safe?'; and 'Is there anything that can and should be done about it?'.

SOURCES OF HELP

Some useful text books and other publications on tree pests, diseases and disorders are listed in the Bibliography (p.299).

Some establishments which employ specialists in tree problems and who offer advisory services are listed in Appendix 2 (p.306).

Hints on collecting and submitting specimens for laboratory examination are given in Appendix 1 (p.305).

TABLE I

THE CAUSES OF ILL-HEALTH IN TREES

LIVING (BIOTIC) AGENTS	NON-LIVING (ABIOTIC) AGENTS	
Fungi	Extremes of weather:	Chemicals, principally:
Bacteria and other single-celled organisms	• spring & autumn frosts	• herbicides
	• winter cold	• de-icing salt & blown
Viruses	• drought	sea salt
Viroids	• sun scorch	• fuel oils in the soil
Biting insects	• wind	• fuel gases in the soil
Sucking insects & mites	• hail	• localized concentrations
Nematodes	• lightning	of toxic fumes in
Birds		the air
Mammals including man		
	Soil conditions:	Other:
	• compaction	• factors associated with
	• surface sealing	transplanting
	• waterlogging	• mechanical agents
	• mineral deficiencies	(e.g. strangulating tree
		ties, abrading stakes,
		impacting vehicles)
		• fire
		• graft failure

RARE AND MISAPPREHENDED CAUSES OF ILL-HEALTH IN TREES

RARE CAUSES OF VISIBLE DAMAGE	PHENOMENA OFTEN THOUGHT TO BE BUT DOUBTFULLY EVER THE CAUSE OF VISIBLE DAMAGE
Cement and other dust deposits	Generalized atmospheric pollution (including acid rain) (iv)
Street Lighting (v)	Vehicle exhaust fumes (iv)
Water-table lowering (iii)	Fuel jettisoned from flying aircraft
	Old age in itself

Note: i   Several agents may operate concurrently or in succession to produce the disease syndrome.

ii   Some disorders, including some common ones, have not yet been fully explained.

iii   See Helliwell, 1993.

iv   See p.97.

v   See p.7.

TABLE II
INFORMATION REQUIRED ON TREE, SITE AND DAMAGE AS THE BASIS FOR
A DIAGNOSIS (see mnemonic on p.16)

INFORMATION TO BE COLLECTED	THE VALUE OF THE INFORMATION
1. The tree	
• Species affected	Immediately narrows the range of possible causes (see Table IV, p.18 and Section 2). A wide range of affected species suggests chemical, site or weather damage; damage to only one species (if others are present) is more suggestive of a pest or disease.
• The tree in its setting	De-icing salt damage is more likely on roadsides than in parkland; site disturbance is more likely on a new housing estate than in woodland; and so on.
• Age or planting date. If planted in previous 2 yrs, type of plant	New plantings often suffer from cultural problems associated with transplanting or after care, especially if stock was bare-rooted or over-large. Verticillium or Phytophthora diseases can be contracted in the nursery.
• Cultural operations in the 2 years preceding damage	May themselves cause damage (e.g. herbicides), or facilitate infections (e.g. pruning and Silver leaf). Poor weed control can severely inhibit establishment.
2. The site	
• Soil type and conditions; previous land use	May be harmful (e.g. liable to waterlogging; very acidic or alkaline), harbour pests or diseases (cockchafers in old grassland; Honey fungus in former woodland), or encourage them (long grass and voles; wet soils and Phytophthora).
• Changes in recent years preceding onset of symptoms	Soil compaction or raised levels can suffocate roots; levelling can damage roots (be suspicious of new buildings, paving or gardens); builders and gardeners often light bonfires; new road repairs may indicate a gas leak.
• Chemicals applied in 12 months preceding onset of symptoms	Herbicides often damage nearby trees: note distortion, discoloration or absence of weeds, peculiar odours. Spillages and deliberate poisonings also occur. Some insecticides and fungicides damage trees. See also Table III, I and 3 (p.14 and p.15) and p.93.

SECTION 1

INFORMATION TO BE COLLECTED	THE VALUE OF THE INFORMATION
3. The problem	
• How widespread?	Some problems are usually localised (root disease, lightning), others widespread (frost, drought, leaf pests and diseases).

Symptoms

• General description	Many problems have distinctive 'signatures': see Table III, p.14 & Section 3.
• Distribution of affected plants and plant parts; pattern of damage on leaves, shoots	Damage may be associated with site features (roads, wet areas, frost hollows). Directional damage suggests wind involvement (wind, hail, sea spray) or bonfire damage or connection with a site feature (roads, pavements). Damaged groups of trees suggests a root problem or lightning. Damaged lower foliage suggests leaf pests or diseases or spring frost; dying upper crown suggests girdling stem damage. See also Table III, p.14.
• Time of onset of symptoms; progress of damage within and between trees	Some conditions are ephemeral (lightning, fire), some recurrent (some foliage pests and diseases), others progressive (some root and bark diseases). The onset of symptoms may coincide with or closely follow a damaging event (drought, building, herbicide application): check annual shoot growth for sudden or progressive reductions; note degree of deterioration of damaged parts; question local residents. Progress of death along a row of trees or as an enlarging group indicates root disease or spread of an infection via root contact.
• Other features associated with the damaged tree	Note any foreign bodies or substances, creatures, outgrowths, abnormalities of growth, odours (sniff excavated soil) or unusual markings: these will include fungal fruit bodies, mycelium or spore masses; bacterial slime; insects, mites or their remains (shed skins, eggs) or signs (frass, waxy wool covering, honey dew, sooty moulds, exit holes); mammals' teeth marks, hair, spoor, droppings; resin, gum or watery fluxes from wounds or dead bark; outgrowths or deformities induced by biotic or abiotic agents (galls etc., petiole or shoot twisting, leaf cupping and so on); abnormal bark coloration. Use a hand lens to examine both sides of damaged leaves.

TABLE III
EXTERNAL SYMPTOMS AS A GUIDE* TO THE SEAT OF DAMAGE

CHARACTERISTIC SYMPTOMS	PART OF TREE SUFFERING PRIMARY DAMAGE
1. Most trees in the vicinity are unaffected	
a. Thin crown, small pale leaves or needles; perhaps also dead twigs, branches or branch ends; OR tree flushes much later and more feebly than usual; or a deciduous tree sheds its leaves much earlier than its neighbours of same species (but see also 2a below).	ROOTS or ROOT COLLAR (killed or function impaired) OR GRAFT UNION (tissues of scion and rootstock imperfectly united or killed at union)
b. One or several scattered branches or top of tree with wilted, faded, dead or yellowed leaves or needles, firmly attached. *If damage is below the live crown, the whole crown will be affected.*	BARK and CAMBIUM of STEM, BRANCHES or TWIGS (killed, removed or constricted)
c. In deciduous broadleaves, one or several scattered branches with wilted, faded, yellowed or dead falling leaves; or with abnormally small leaves.	WATER-CONDUCTING WOOD (infected by wilt-disease fungi or carrying certain toxic chemicals).
d. A snapped branch or stem; or fruit bodies of wood-decaying fungi.	STRUCTURAL WOOD (decayed - perhaps at some distance from fruit bodies)
2. Most trees of the same species in the vicinity are affected	
a. Young or old, irregularly spotted blotched or mottled leaves; banded or stippled needles; or a white or dark coating of mould or similar substance on foliage (rubs off with thumb). Commonly, such needles or leaves are of one age and fall prematurely (but see also 1a above). The damage may decrease in severity with height up the tree.	FOLIAGE (diseased, or infested with sucking insects or mites or leaf miners).
b. Yellowing (and perhaps browning) of older leaves or needles (i.e. in evergreens, those produced before the current year; in deciduous species, those towards the shoot base) which then fall prematurely. Remaining (broad) leaves may be wilted.	FOLIAGE (damaged by drought)

CHARACTERISTIC SYMPTOMS	PART OF TREE SUFFERING PRIMARY DAMAGE

3. Other trees in the vicinity may or may not be affected

In broadleaves, a symmetrical pattern of foliar discoloration or necrosis e.g. interveinal, veinal or marginal. In conifers, a uniform or evenly graduated discoloration of needles, e.g. from needle tips back without a sudden transition from one colour to another; or a uniform discoloration of all needles on a shoot. Symmetry is also evident in the age of leaves or needles affected, e.g. all ages, or only current ones, or only older ones.	FOLIAGE (damaged by certain toxic chemicals, e.g. some herbicides, or by a lack of certain nutritive minerals such as iron, magnesium).

*i   The symptoms listed here are typically but not invariably induced by damage to the part of the tree stated.

ii   Once a tree is virtually dead, external symptoms are of much less diagnostic value.

DIAGNOSIS: A FIELD PROCEDURE

The FLOW CHART and the associated Tables I – IV, if used in conjunction with the descriptions of diseases and disorders in Section 3, offer the reader, confronted with an ailing tree, an orderly way of investigating the problem and a fair chance of reaching a sound diagnosis. It does not guarantee success every time – often the problem will need the help of specialists or will be insoluble – but it will lead the investigator along profitable lines of inquiry so that any diagnosis, however tentative, will be based on good evidence and sound reasoning.

For the beginner, it should illuminate the often puzzling and sometimes daunting world of diagnosis; for the more experienced, it is offered as a means of refining working practice and improving the reliability of diagnoses made.

USING THE DIAGNOSTIC FLOW CHART

The chart is intended to be used in conjunction with the information in the rest of this handbook. Before beginning, the whole of Section 1 should be studied.

Step 1.   Start at Box 1. Follow the instructions in the box, then follow the arrow to the next box.

Step 2.   Repeat the process until Box 19, Diagnosis, or Box 18, Problem Unexplained, is reached. Where two arrows leave a box, read the note against each and follow the appropriate route.

Symptoms of ill-health in trees are rarely so distinctive that their cause is self-evident. As similar symptoms can have many different causes, it is usually unsafe to make a diagnosis before all the symptoms and circumstances of the case have been considered.

A SUMMARY OF THE INFORMATION REQUIRED FOR A DIAGNOSIS – A MNEMONIC (see Table II p.12-13)

D istribution of damage

I dentity of tree

S ite type, conditions and changes

E nvironmental conditions and other changes (e.g. weather, cultural treatments, chemicals applied)

A ge of tree now and/or at planting

S ymptoms

E xtraneous matter (e.g. insects, fungi, exudates, odours)

D ead bark, **D**iscoloured wood – check by cutting

T ime of onset of damage and its progress with time

R oot condition

E vidence – review before making a diagnosis

E xtra evidence – seek where possible, in order to confirm your diagnosis

A DIAGNOSTIC PROCEDURE

- Start at box 1 and follow the appropriate arrows.
- To avoid a wrong or incomplete diagnosis, do not omit any steps.
- Remember that several different problems may be evident at once.
- For more information, see p.16

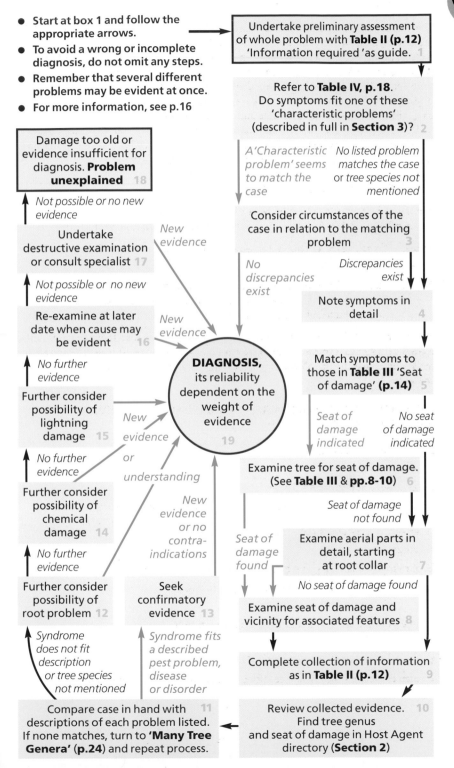

1 Undertake preliminary assessment of whole problem with **Table II (p.12)** 'Information required 'as guide.

2 Refer to **Table IV, p.18**. Do symptoms fit one of these 'characteristic problems' (described in full in **Section 3**)?

A 'Characteristic problem' seems to match the case

No listed problem matches the case or tree species not mentioned

3 Consider circumstances of the case in relation to the matching problem

No discrepancies exist

Discrepancies exist

4 Note symptoms in detail

5 Match symptoms to those in **Table III** 'Seat of damage' **(p.14)**

Seat of damage indicated

No seat of damage indicated

6 Examine tree for seat of damage. (See **Table III** & **pp.8-10**)

Seat of damage not found

Seat of damage found

7 Examine aerial parts in detail, starting at root collar

No seat of damage found

8 Examine seat of damage and vicinity for associated features

9 Complete collection of information as in **Table II (p.12)**

10 Review collected evidence. Find tree genus and seat of damage in Host Agent directory (**Section 2**)

11 Compare case in hand with descriptions of each problem listed. If none matches, turn to **'Many Tree Genera' (p.24)** and repeat process.

Syndrome does not fit description or tree species not mentioned

Syndrome fits a described pest problem, disease or disorder

13 Seek confirmatory evidence

12 Further consider possibility of root problem

No further evidence

14 Further consider possibility of chemical damage

No further evidence

15 Further consider possibility of lightning damage

No further evidence

16 Re-examine at later date when cause may be evident

New evidence

Not possible or no new evidence

17 Undertake destructive examination or consult specialist

New evidence

Not possible or no new evidence

18 Damage too old or evidence insufficient for diagnosis. **Problem unexplained**

19 DIAGNOSIS, its reliability dependent on the weight of evidence

New evidence or understanding

New evidence or no contra-indications

TABLE IV
PROBLEMS OF ILL HEALTH CHARACTERISTIC OF SOME COMMON AMENITY TRE[ES]

	Honey fungus (p.145)	Phytophthora Root Disease (p.206)	Other Root Diseases (Sec. 2)	Verticillium Wilt (p.261)	Fireblight (Sec. 2)	Nectria Canker (p.124)	Silver Leaf (p.185)	Leaf & Shoot Aphids; Mites etc. (p.252)	Leaf Miners (p.158)	Other problems
Acer	✓			✓				✓		
Aesculus	✓									Guignardia leaf blotch (p.141); Bleeding canker (p.202)
Araucaria	✓									
Betula	✓		✓					✓		Leaf rusts (p.241)
Castanea		✓								
Catalpa				✓						
Cedrus	✓		✓					✓		
Chamaecyparis	✓	✓						✓		
Cotoneaster					✓					
Crataegus		✓	✓		✓					
xCupressocyparis	✓							✓		Coryneum canker (p.107)
Cupressus								✓		Coryneum canker (p.107)
Fagus		✓	✓			✓		✓	✓	
Fraxinus						✓				Bacterial canker (p.69); Bud moth (p.68)
Ilex									✓	
Juglans	✓									Marssonina leaf spot (p.167)

SECTION 1

p.9)

Genus	Phomopsis canker (p.201)	Apple scab (p.245)	Top dying (p.47)	Needle cast diseases (p.189–190)	Anthracnose (p.166)	Bacterial canker (p.70); Leaf rusts (p.241)	Cherry leaf scorch (p.66), Blossom wilt (p.85)	Anthracnose (p.167); Willow scab (p.245)	Winter cold browning (p.129)	Lightning (p.170)	Unexplained branch death (p.277)	Scale insects (p.149)	Dutch elm disease (p.116)
Juniperus	✓												
Laburnum											✓		
Liriodendron											✓		
Malus		✓											
Picea			✓	✓									
Pinus				✓									
Platanus					✓								
Populus						✓							
Prunus						✓	✓						
Quercus										✓			
Salix						✓		✓					
Sequoia									✓				
Sequoiadendron										✓			
Sorbus (Rowans)		✓											
Sorbus (Whitebeam)		✓											
Taxus											✓		
Thuja	✓												
Tilia												✓	
Ulmus												✓	✓

Caution:
i Blanks do not indicate immunity.
ii Not all members of a genus are subject to the problems indicated.
iii Always check that the observed symptoms and circumstances tally with a reliable published description of the suspected disease or disorder before making a final diagnosis.
iv Honey fungus is uncommon in street trees.

SECTION 2

A HOST-AGENT DIRECTORY

ALPHABETICALLY ARRANGED BY TREE GENUS OF
THE PESTS, DISEASES, DISORDERS AND MISCELLANEOUS
PHENOMENA DESCRIBED IN THIS BOOK

EXPLANATORY NOTES

N.B. In this Directory, trees are listed in alphabetical order by their Latin names. English names with their Latin equivalents are indexed at the end of the book.

To use this Host-Agent Directory effectively, the investigator needs to know the species (or at least the genus) of the tree in question, to have ascertained (or at least to have carefully considered) the location of the seat of damage as described in Section 1, and to have collected fairly detailed information on the symptoms and circumstances of the case (also described in Section 1).

Thus armed, **turn to the appropriate tree genus page** (or pages if trees of different genera are involved).

Under each tree genus are listed a number of headings (except where no useful information on the pathology of that particular genus has been found). These are 'Seats of damage', i.e. the part of the tree on which the damaging agent is having a direct effect. **Consult the list under the appropriate Seat of Damage**.

These lists give the types of damage and agents of damage most often encountered on the tree genus in question. Whether damage type or agent is given depends on which is most likely to be the subject of the inquiry. The damage types are given either in a helpful descriptive phrase (Leaf galls, Phytophthora root killing, Nectria canker) or as 'Damage' caused by a certain agent (e.g. Salt damage, Lightning damage, Leaf hopper damage). In the latter case, the type of damage caused cannot be usefully summarized in a short phrase but is described on the page given.

Turn to the page given against the type or agent of damage which seems most like the problem under consideration (or to all in turn if none seems to match). These page numbers refer to the full descriptions of pests, diseases and disorders in Section 3. **Compare the information given in the description with the problem in hand**. This may allow a diagnosis to be made, either directly or after further confirmatory evidence has been sought.

If none of the suggested causes listed for the species of tree in question fits the problem under consideration, turn to the first page of the Host-Agent Directory, 'Many Tree Genera', and repeat the process described above.

If a diagnosis still cannot be made, return to the Flow Chart and follow the instructions there.

It should be noted that this Directory is not an exhaustive list of all that might be encountered on each genus: see 'The pests, diseases and disorders included' on page x.

MANY TREE GENERA

■ **Roots, root collar, butt**

Honey fungus	145
Phytophthora root disease	206
Root rots*: *Ganoderma adspersum,*	
G. applanatum, Het. annosum	226
Coatings, deposits, exudations	
and outgrowths	100
Waterlogging	266
Mechanical injury	176
Galls	132

■ **Bark, cambium of stem, branches, twigs**

Coral spot fungus	106
Scale insects	248
Squirrel damage	178
Lightning damage	170
Mechanical injury	176
Fire damage	122
Chemical damage	93
Drought damage	112
Lichens	100
Algae	100
Athelia	100
Woolly aphids	273,275
Coatings, deposits, exudations	
and outgrowths	100
Galls	132
see also Asian longhorn beetle†	281

■ **Foliage (leaves, shoots, buds), flowers, fruit**

Leaf spots and blotches	161
Powdery mildews†	215
Virus diseases†	263
Leaf mines	158
Leafhopper damage†	157
Scale insects	248
Galls	132
Leopard moth damage†	169
Honeydew and sooty moulds	144
Slugworm damage†	195,198
Lime induced chlorosis	174
Drought damage	112
Frost damage	126,129
Winter cold damage	129
Lightning damage	170

Wind damage	176
Fire damage	122
Chemical damage	93
Mechanical injury	176
Coatings, deposits, exudations	
and outgrowths	100
Nutrient deficiencies	176
Galls	132
Street lighting/autumn frost	
damage	130

■ **Structural and water-conducting wood**

Stem and top rots*: *Ganoderma*	
adspersum, G. applanatum,	
Coriolus versicolor	226
Verticillium wilt†	261
Leopard moth damage (in branches	
up to c. 10 cm diameter; rarely,	
larger stems are affected similarly	
by Goat moth†	169
Bacterial wetwood	74
see also Asian longhorn beetle†	281

* Many species of decay fungi occasionally occur on trees other than their usual host species.
† = on broadleaved trees only

ABIES – Silver firs

■ **Roots, root collar, butt**

Phytophthora Killing	206

■ **Bark, cambium of stem, branches, twigs**

Adelgid wool	76,251

■ **Foliage (leaves, shoots, buds), flowers, fruit**

Distortion from	
adelgid feeding	76,251,273
Honeydew and Sooty	
moulds	144
Woolly aphids	251,273
Needle mines	160
Salt damage	93
Winter cold damage	129
see also Siberian fir woolly aphid	282

- **Structural and water-conducting wood**

  Drought crack   112

ACACIA – Mimosa

- **Foliage (leaves, shoots, buds), flowers, fruit**

  Winter cold damage   129
  Chalk chlorosis   174

- **Structural and water-conducting wood**

  Drought crack   112
  Watery exudate from pruning
  or other wounds   103

Comments

  [Roots: *Heterobasidion annosum* not recorded – see p.220]

ACER – Maples

- **Roots, root collar, butt**

  Honey fungus killing (see also
  Comment (1) below)   145
  Root and butt rots: *Ganoderma
  adspersum, G. applanatum, Ustulina
  deusta*   226

- **Bark, cambium of stem, branches, twigs**

  Nectria canker   185
  Coral spot fungus   106
  Squirrel damage   178
  Scale insects   149
  Drought damage   112
  Winged cork   103

  See also Sooty bark disease   257

- **Foliage (leaves, shoots, buds), flowers**

  Leaf spots and blotches   162,163
  Powdery mildew   215
  Leaf galls   133,134
  Leafhopper damage   157
  Honeydew and Sooty moulds   144

  Drought damage   112
  Salt damage   93

  See also Comment (2) below

- **Structural and water-conducting wood**

  Top rot: *Bjerkandera adusta*   226
  Verticillium wilt   261
  Sooty bark disease   257
  Leopard moth damage (twigs
  and small branches)   169
  Watery exudate from pruning
  and other wounds   103

Fig. 1 *Leaf scorch of sycamore*

Comments

1. The lack of records of Honey fungus on *Acer negundo*, even in its North American home, suggests that it may be immune to the disease.

2. The cause of marginal and inter-veinal leaf browning of *Acer* ('leaf scorch') (Fig. 1), quite common on *A. pseudoplatanus* and *A. ps. 'Brilliantissimum'* and not caused by toxic chemicals or drought, is not known

AESCULUS – Horse chestnuts

- **Roots, root collar, butt**

  Honey fungus killing   145
  Phytophthora killing   206

For problems affecting many tree genera, see p.24

Root and butt rots: *Armillaria,*
Ganoderma adspersum, G.
applanatum, Ustulina deusta   226

■ **Bark, cambium of stem,**
branches, twigs

Phytophthora bleeding canker   202
Scale insects   149
Squirrel damage   178
Galls (Bud Proliferation)   89

■ **Foliage (leaves, shoots, buds),**
flowers, fruit

Guignardia leaf blotch   141
Leaf tatter   Fig.135
Leafhopper damage   157
Scale insects   149
Leopard moth damage   169
Squirrel damage to shoots   179
Bud proliferation   89
Salt damage   93

See also Comments below

■ **Structural and water-conducting**
wood

Top rots: *Ganoderma adspersum,*
G. applanatum, Pleurotus ostreatus 226
Bacterial wetwood   74
Leopard moth damage (twigs
and small branches)   169
Wind breakage of branches   176
Twig shedding – see Comment (3)
under *Populus*

Fig. 2 *Leaf scorch of Horse Chestnut*

Comments

The cause of a marginal leaf browning ('leaf scorch'), recurrent on certain Horse chestnut trees, has not
been established (Fig.2). The possibility that Xylem limited bacteria
(p.5) are involved merits further
research.

AILANTHUS – Trees of Heaven

■ **Roots, root collar, butt**

Root and butt rot:
Pholiota squarrosa   220

■ **Foliage (leaves, shoots, buds),**
flowers, fruit

Twig shedding – see Comment (3)
under *Populus*

■ **Structural and water-conducting**
wood

Top rots: *Laetiporus sulphureus,*
Bjerkandera adusta,   226
Verticillium wilt   261

Comments

[Roots: *Heterobasidion annosum* not
recorded – see p.220]

ALNUS – Alders

■ **Roots, root collar, butt**

Phytophthora killing   205

■ **Bark, cambium of stem,**
branches, twigs

Scale insects   250,251

See also Comment 1 below

■ **Foliage (leaves, shoots, buds),**
flowers, fruit

Taphrina leaf distortion and
thickening   197
Leaf galls   133,134

Leaf mines   158
Leaf beetle damage   156
Crown thinning bud death, - See
Comment 2 below

Comments

1. The cause of a progressive and sometimes fatal dieback disease common
and widespread on *A. glutinosa* in
highland Scotland and noted at a
few sites on this and on *A. incana* in
orchard wind-breaks further south
has not been established. It appears
to be the result of a pathogen
spreading in the bark and cambium,
and perhaps the wood, from the
branches down the stem into the
roots. Dead strips of bark may be
visible on living stems of diseased
trees as long, narrow depressions
2. Bud death during late winter
(March) leading to a thinning of the
crown foliage can be caused by larvae of the bud-mining tortricid
moth *Epinotia tenerana* (see Gregory,
MacAskill & Winter 1996).

AMELANCHIER – Snowy Mespils

- **Bark, cambium of stem,
branches, twigs**

  Fireblight damage   124

- **Foliage (leaves, shoots, buds),
flowers, fruit**

  Leaf mines   158
  Lime induced chlorosis   174

Comments

[Roots: *Heterobasidion annosum*
recorded only on the shrubby
A. spicata – see p.220]

ARAUCARIA – Monkey Puzzles

- **Roots, root collar, butt**

  Honey fungus killing   145
  Phytophthora killing   206

Root and butt rots: *Armillaria,
Ganoderma adspersum* or
applanatum   226

- **Foliage (leaves, shoots, buds),
flowers, fruit**

  Air pollution damage   96
  Salt damage   93

ARBUTUS – Strawberry Trees

- **Roots, root collar, butt**

  Phytophthora killing   206

- **Bark, cambium of stem,
branches, twigs**

  See Comments below

Comments

The cause of a perennating stem
canker has not been established but
the possible involvement of
Phytophthora cactorum merits
research as this causes a basal canker
on *Arbutus* in the USA and the aerial
Bleeding canker disease on *Aesculus*
in this country.

ATHROTAXIS – Tasmanian Cedars

Comments

[Roots: *Heterobasidion annosum* not
recorded – see p.220]. The authors
have found no other useful information on the pathology of *Athrotaxis*.

AUSTROCEDRUS –
Southern Incense Cedars

Comments

[Roots: *Heterobasidion annosum* not
recorded – see p.220]. The authors
have found no other useful information on the pathology of
Austrocedrus.

SECTION 2

BETULA – Birches

■ **Roots, root collar, butt**

Honey fungus killing	145
Phytophthora killing	206
Root and butt rots: *Armillaria*,	
Heterobasidion annosum	220

■ **Bark, cambium of stem, branches, twigs**

Scale insects	248

■ **Foliage (leaves, shoots, buds), flowers, fruit**

Leaf spots	163
Leaf rust fungi	241
Leaf galls	134
Leaf mines	158
Leafhopper damage	157
Bud galls	135
Slugworm damage	195
Honeydew and Sooty moulds	144
Drought damage	112
Salt damage	93

■ **Structural and water-conducting wood**

Watery exudate from pruning and other wounds	103
Top rots: *Piptoporus betulinus*, *Daldinia concentrica*, *Inonotus obliquus*, *Fomes fomentarius*	226

See also Comment below

■ **Other**

Witches' brooms	270

Comment

The pathogenic ability of *P. betulinus* and, in the North, of *I. obliquus* and *F. fomentarius*, so often seen on ailing birch trees, is uncertain, but they are probably unable to progress far in trees unless these have first been damaged or stressed by such things as storm, drought or waterlogging. *P. betulinus* may well be an endo-

phytic pathogen (q.v.). See Gregory and Redfern, 1998.

BUXUS – Box

■ **Bark, cambium of stem, branches, twigs**

Volutella leaf and twig blight	265

■ **Foliage (leaves, shoots, buds), flowers, fruit**

Rust-thickened leaves	243
Volutella leaf and twig blight	265
Distortion (galls) from psyllid feeding	86
Psyllid wool	86
Late (spring) frost damage – see Comments below	126
Salt damage	93

Fig. 3 *Cylindrocladium blight of box.*

Comments

1. Spring frost damage takes the form of leaf bleaching – see Fig. 141 [Roots: *Heterobasidion annosum* not recorded – see p.220]
2. A blight of *Buxus sempervirens* in which leaves become pale brown, shrivelled and fall, and black streaks develop on affected stems, and large patches of defoliated shoots die (Fig. 3), was troubling some growers in

For problems affecting many tree genera, see p.24

the south of England in the late 1990s. At the time of writing the cause had not been demonstrated conclusively but a fungus in the genus *Cylindrocladium* is consistently associated with the damage. Control consists in destroying infected plants or plant parts. Fallen infected leaves are harmless if buried. At the time of writing, no fungicidal control had been developed.

CALOCEDRUS – Northern Incense Cedars

■ **Foliage (leaves, shoots, buds), flowers, fruit**

Scale insects	248

CARPINUS – Hornbeams

■ **Bark, cambium of stem, branches, twigs**

Coral spot fungus	106
Scale insects	248

■ **Foliage (leaves, shoots, buds), flowers, fruit**

Leaf spots	164
Leaf galls	132
Leafhopper damage	157
Leaf mines	158
Salt damage	93

■ **Structural and water-conducting wood**

Watery exudate from pruning and other wounds	103

■ **Other**

Witches' brooms	270

CARYA – Hickories

■ **Roots, root collar, butt**

Waterlogging	266

■ **Foliage (leaves, shoots, buds), flowers, fruit**

Salt damage	93
White mould	165

CASTANEA – Chestnuts

■ **Roots, root collar, butt**

Phytophthora killing (Ink disease)	206

Root and butt rots:
Laetiporus sulphureus, Hypholoma fasciculare, Fistulina hepatica　226

■ **Bark, cambium of stem, branches, twigs**

see also Chestnut blight	280

■ **Foliage (leaves, shoots, buds), flowers, fruit**

Leaf mines	158
Lime induced chlorosis	174

Comments

[Roots: *Heterobasidion annosum* not recorded – see p.220]

CATALPA – Catalpas

■ **Roots, root collar, butt**

Root and butt rot:
Pholiota squarrosa　220

■ **Foliage (leaves, shoots, buds), flowers, fruit**

Twig shedding – see Comment (3) under *Populus*

■ **Structural and water-conducting wood**

Verticillium wilt	261

Comments

[Roots: *Heterobasidion annosum* not recorded – see p.220]

SECTION 2

CEDRELA – Chinese Cedar

Comments

[Roots: *Heterobasidion annosum* not recorded – see p.220]. The authors have found no other useful information on the pathology of *Cedrela*.

CEDRUS – Cedars

■ **Roots, root collar, butt**

Honey fungus killing	145
Root and butt rots: *Phaeolus schweinitzii, Sparassis crispa, Armillaria, Coniophora puteana* (butt rot)	226

■ **Bark, cambium of stem, branches, twigs**

Coniophora puteana	220
Phacidium branch killing. See Comment (1) below	
See also Comment (2) below	

■ **Foliage (leaves, shoots, buds), flowers, fruit**

Aphid damage	254
Honeydew and Sooty moulds	144
Hormone herbicide damage	93

Fig. 4 *Cedar branches killed by the fungus Phacidium coniferarum*

Comments

1. Dead and dying twigs and small branches are quite common on Cedrus (Fig.4). In the few cases investigated, dying, girdling bark lesions have usually yielded the fungus *Phacidium coniferarum* (anamorph: *Phacidiopycnis pseudotsugae*). This is a well known cause of such damage on *Pseudotsuga* and, less often, *Larix* in forest nurseries and plantations. For appearances' sake, dead parts can be removed, but spread ceases before large branches are girdled. When first described in 1920 (from Scotland), the fungus was mistakenly thought to be a *Phomopsis* species and named *Ph. pseudotsugae* (see Peace 1962; Smith et al 1988).

2. A slow, true dieback of odd, large branches of *Cedrus deodara* has been observed on several occasions in England but remains unexplained. It has the hallmarks of a spreading, perennating fungal bark infection but inoculations with the only fungus frequently isolated from the dying bark (*Placosphaera corrugata*) have not induced the disease.

CELTIS – Nettle Trees

■ **Foliage (leaves, shoots, buds), flowers, fruit**

Salt damage	93

Comments

[Roots: *Heterobasidion annosum* not recorded – see p.220].

CEPHALOTAXUS – Cow's Tail Pines

■ **Foliage (leaves, shoots, buds), flowers, fruit**

Artichoke gall	279

Comments

[Roots: *Heterobasidion annosum* not recorded – see p.220].

CERCIDIPHYLLUM – Katsura Tree

■ **Foliage (leaves, shoots, buds), flowers, fruit**

Late (spring) frost   126

Comments

[Roots: *Heterobasidion annosum* not recorded – see p.220]

CERCIS – Judas Trees

■ **Bark, cambium of stem, branches, twigs**

Coral spot fungus   106

■ **Structural and water-conducting wood**

Verticillium wilt   261

CHAMAECYPARIS – False Cypresses

■ **Roots, root collar, butt**

Honey fungus killing   145
Phytophthora killing   206

■ **Bark, cambium of stem, branches, twigs**

Amylostereum canker rot   64

■ **Foliage (leaves, shoots, buds), flowers, fruit**

Kabatina shoot blight   152
Aphid damage   109
Scale insects   249
Mined shoots   151
Drought damage   112
Salt damage   93
Urine damage   93
Magnesium deficiency   114

■ **Structural and water-conducting wood**

Amylostereum canker rot   64

CHRYSOLEPIS – Golden Chestnut

Comments

The authors have found no useful information on the pathology of *Chrysolepis*

CLADRASTIS – Yellow Woods

Comments

[Roots: *Heterobasidion annosum* not recorded – see p.220]. The authors have found no other useful information on the pathology of *Cladrastis*.

CORDYLINE – Cabbage Tree

■ **Bark, cambium of stem, branches, twigs**

Winter cold damage   129

■ **Foliage (leaves, shoots, buds), flowers, fruit**

Winter cold damage   129

Comments

[Roots: *Heterobasidion annosum* not recorded – see p.220].

CORNUS – Dogwoods

■ **Roots, root collar, butt**

Waterlogging   266

See Comment (1) below

■ **Bark, cambium of stem, branches, twigs**

Scale insects   149

See also Comments (1) and (2) below

■ **Foliage (leaves, shoots, buds), flowers, fruit**

SECTION **2**

Powdery mildew　　　　　215
See also Comment (2) below

Fig. 5 *Anthracnose of flowering dogwood*

Comments

1. In North America, where it is native, *Cornus nuttallii* is subject to a root collar canker caused by *Phytophthora cactorum* (cf Bleeding canker on p.202) and to Nectria canker (p.185).
2. In the late 1990s the death of foliage and branches of *C.florida* was reported from several widely separated places in southern England. Brown spots and blotches appear on flowers, leaves (Fig. 5) and young shoots and dieback of branches ensues. Epicormic recovery shoots may develop only to be killed in turn. Cankers develop on the branches and main stem. Damage often begins in the lower crown and spreads upwards. The disease is probably the same as that caused in the USA by *Discula destructiva* and known there as Anthracnose of flowering dogwood, but the precise identity of the fungus and its origins are still being investigated on both sides of the Atlantic. In America, *Cornus* species and varieties vary widely in susceptibility, from the very susceptible *C.florida*, *C.kousa* var. *chinensis* and *C.nuttallii* to the very resistant *C.kousa* and *C.ma*s. Control with fungicides such as chlorthalonil or propiconazole is possible but repeated, severe damage is best dealt with by replanting with resistant species.

CORYLUS – Hazels

■ **Roots, root collar, butt**

Honey fungus killing and rot　　145

■ **Foliage (leaves, shoots, buds), flowers, fruit**

Powdery mildew　　　　　　215
Leaf mines　　　　　　　　158
Leafhopper damage　　　　　157
Bud galls　　　　　　　　　135

See also Comments below

■ **Structural and water-conducting wood**

Top rot: *Stereum rugosum*　　226

Comments

Small yellow-green spots on leaves and husks, and the death of buds, shoots and branches have been reported on *C. avellana* from the south of England, caused by the bacterium *Xanthomonas campestris pv. corylina* but its significance on amenity trees and its full distribution is not known.

COTONEASTER FRIGIDUS – Cotoneaster

■ **Bark and cambium of stem, branches, twigs**

Fireblight killing　　　　　124
Woolly aphids　　　　　　　275

■ **Foliage (leaves, shoots, buds), flowers, fruit**

Fireblight killing　　　　　124

CRATAEGUS – Thorns

■ **Roots, root collar, butt**

Black fleck disease	81
Phytophthora root killing	206
Root rot: *Heterobasidion annosum*	220

See also Comment (1) below

■ **Bark and cambium of stem, branches, twigs**

Fireblight killing	124
Scale insects	248
Woolly aphids	275

■ **Foliage (leaves, shoots, buds), flowers, fruit**

Leaf spot diseases	164
Leaf rust fungi	244
Powdery mildew	215
Fireblight killing	124
Leaf galls	133,134
Leaf mines	158
Leafhopper damage	157
Slugworm damage	198
Monilia blight	182

See also Comment (2) below

■ **Structural and water-conducting wood**

Mistletoe	103

Comments

1. The scarcity of reported cases of Honey fungus killing on a genus as commonly cultivated as *Crataegus* indicates its very considerable resistance to the disease.

2. Leaves of *Crataegus* are sometimes bronzed by the Rust mite, *Epitrimerus piri*. No control is required.

3. Scattered twigs or branches or whole crowns of *C. monogyna* or *C. oxyacantha* may be debilitated or killed as the result of feeding by the buprestid beetle, *Agrilus sinuatus*, in the phloem and cambium. The

Fig. 6 *Buprestid beetle gallery in hawthorn bark*

foliage of infested branches is yellow, undersized, sparse or lacking. Branches bearing live but symptomatic leaves are alive but towards their base (and also under dead patches of bark on twigs bearing dead leaves) remarkably sinuous beetle galleries can be found in the phloem and cambium, perhaps but not necessarily girdling (Fig. 6). D-shaped exit holes may be present in bark over-lying galleries. The damage may worsen over two or three years, gradually involving more of the crown and more trees, and whole trees may die from multiple infestations of branches and main stems.

Only a handful of cases have been identified since the problem was first noted in London in 1992, all in the southern half of England, but this is almost certainly a reflection of its unfamiliarity and not of its rarity. The timing of the few cases studied strongly suggests that the attacks were all on trees subjected to lengthy periods of drought; the insect is not considered to be capable of attacking unstressed trees.

× CRATAEMESPILUS

■ **Roots, root collar, butt**

Root and butt rot: *Ganoderma species*	226

For problems affecting many tree genera, see p.24

■ **Bark and cambium of stem, branches, twigs**

Fireblight killing 124
Nectria canker 185

■ **Foliage (leaves, shoots, buds), flowers, fruit**

Monilia blight 182

CRYPTOMERIA – Japanese Cedar

■ **Roots, root collar, butt**

Honey fungus killing and rot 145

■ **Foliage (leaves, shoots, buds), flowers, fruit**

Foliage bronzing 130

CUNNINGHAMIA – Chinese Firs

■ **Roots, root collar, butt**

Honey fungus killing and rot 145

■ **Bark and cambium of stem, branches, twigs**

Early (autumn) frost damage 129

Comments

[Roots: *Heterobasidion annosum* not recorded – see p.220]

× CUPRESSOCYPARIS – Leyland Cypress

■ **Roots, root collar, butt**

Honey fungus killing 145
Root and butt rot: *Heterobasidion annosum, Armillaria* 226

■ **Bark and cambium of stem, branches, twigs**

Coryneum canker 107

■ **Foliage (leaves, shoots, buds), flowers, fruit**

Kabatina shoot blight 152

Aphid damage 109
Mined shoots 151
Scale insects 249
Drought damage 112
Honeydew and Sooty moulds 144
Lime induced chlorosis 174

CUPRESSUS – True Cypresses

■ **Roots, root collar, butt**

Honey fungus killing 145
Phytophthora killing 206

■ **Bark and cambium of stem, branches, twigs**

Coryneum canker 107
Phomopsis canker 201
Winter cold damage 129

■ **Foliage (leaves, shoots, buds), flowers, fruit**

Kabatina shoot blight 152
Aphid damage 109
Scale insects 249
Winter cold 129
Honeydew and Sooty moulds 144

■ **Structural and water-conducting wood**

Wood decay: *Leptoporus ellipsosporus* (see Comments below)

Comments

Little is known about the wood rots of *Cupressus* in this country but *Leptoporus ellipsosporus* has been found associated with a distinctive 'ring rot' of *C. macrocarpa*. The fruit bodies of this polypore are small, brown brackets (Fig.326).

CYDONIA – Quince

■ **Bark and cambium of stem, branches, twigs**

Fireblight 124
Woolly aphids 275
Scale insects 248

For problems affecting many tree genera, see p.24

- **Water-conducting wood**

  Verticillium wilt   261

- **Foliage (leaves, shoots, buds), flowers, fruit**

  Leaf spots   164
  Powdery mildew   215
  Monilia blight   181
  Slugworm damage   198

Comments

[Roots: *Heterobasidion annosum* not recorded – see p.220]

DAVIDIA – Dove, Ghost or Handkerchief Tree

- **Foliage (leaves, shoots, buds), flowers, fruit**

  Late (spring) frost damage   126
  Drought damage   112

Comments

[Roots: *Heterobasidion annosum* not recorded – see p.220]

DIOSPYROS – Persimmons

- **Foliage (leaves, shoots, buds), flowers, fruit**

  Powdery mildew   215

DRIMYS – Winter's Bark

- **Bark and cambium of stem, branches, twigs**

  Winter cold damage   129

- **Foliage (leaves, shoots, buds), flowers, fruit**

  Winter cold damage   129

Comments

[Roots: *Heterobasidion annosum* not recorded – see p.220]

EMBOTHRIUM – Chilean Firebush

- **Roots, root collar, butt**

  Honey fungus killing   145

- **Bark and cambium of stem, branches, twigs**

  Winter cold damage   129

- **Foliage (leaves, shoots, buds) flowers, fruit**

  Lime induced chlorosis   174
  Winter cold damage   129

Comments

[Roots: *Heterobasidion annosum* not recorded – see p.220]

EUCALYPTUS – Gums

- **Roots, root collar, butt**

  Phytophthora root killing   206

- **Bark and cambium of stem, branches, twigs**

  Silver leaf killing   252
  Winter cold damage   129

- **Structural and water-conducting wood**

  Silver leaf staining   252
  see also Eucalyptus borer   282

- **Foliage (leaves, shoots, buds), flowers, fruit**

  Winter cold damage   129

EUCRYPHIA – Eucryphia

- **Foliage (leaves, shoots, buds), flowers, fruit**

  Winter cold damage   129
  Lime induced chlorosis   174

SECTION 2

Comments

[Roots: *Heterobasidion annosum* not recorded – see p.220]

FAGUS – Beeches

■ **Roots, root collar, butt**

Root, butt rots: *Armillaria,*
Meripilus giganteus, Ganoderma
adspersum, G. applanatum,
G. pfeifferi, Heterobasidion
annosum, Ustulina deusta   226
Phytophthora killing   206

■ **Bark and cambium of stem,**
branches, twigs

Beech bark disease   77
Nectria canker   185
Squirrel damage   178
Stereum canker rot   259
Coral spot fungus   106
Ascodichaena fungus   104
Quaternaria fungus   102
Felted beech scale (coccus)   120
Drought damage   112
Strip cankers   114

■ **Foliage (leaves, shoots, buds),**
flowers, fruit

Leaf spots and blotches   164
Leaf mines   79
Woolly aphids   80
Leaf galls   133,134
Leafhopper damage   157
Slugworm damage   195
Lime induced chlorosis   174
Drought damage   112
Salt damage   93
Honeydew and Sooty moulds   80

See also Comment below

■ **Structural wood**

Top rots: *Bjerkandera adusta,*
Coriolus versicolor, Chondrostereum
purpureum, Ganoderma adspersum,
G. applanatum, G. pfeifferi,
Hypoxylon spp., Pleurotus

ostreatus, Polyporus squamosus,
Pseudotrametes gibbosa   226

Comments

Young *Fagus sylvatica*, like *Quercus robur*, often retain many of their dead leaves over winter. This is not a pathological condition (Fig.15).

FICUS CARICA – Fig

■ **Roots, root collar, butt**

Honey fungus killing   145

■ **Bark and cambium of stem,**
branches, twigs

Coral spot fungus   106
Diaporthe canker – see Comments below
Winter cold damage   129

Comments

Twigs and branches may be cankered or girdled and killed by the fungus *Diaporthe cinerascens* (anamorph: *Phomopsis cinerascens*). This enters through pruning and other bark wounds. Infection can be prevented if fresh wounds are treated with a fungicidal canker paint. Perennating cankers can be prevented from girdling branches if the infected bark is excised and the wound treated with the canker paint. It was first studied in Italy in the 19th Century and has been known in Great Britain since the early nineteen hundreds.

[Roots: *Heterobasidion annosum* not recorded – see p.220]

FITZROYA – Patagonian Cypress

Comments

[Roots: *Heterobasidion annosum* not recorded – see p.220]. The authors have found no other useful information on the pathology of *Fitzroya*.

For problems affecting many tree genera, see p.24

FRAXINUS – Ashes

■ **Roots, root collar, butt**

Root and butt rots: *Ganoderma adspersum, G. applanatum, Perenniporia fraxinea, Pholiota squarrosa*   226

■ **Bark and cambium of stem, branches, twigs**

Bacterial canker	69
Nectria canker	185
Scale insects	248,250

See also Comments (1) and (2) below

■ **Foliage (leaves, shoots, buds), flowers, fruit**

Ash bud moth damage	68
Insect and mite galls	133,134,135

See also Comments below

■ **Structural and water-conducting wood**

Top rots: *Daldinia concentrica, Inonotus hispidus*   226
Verticillium wilt see Comment (3) below

Fig. 7 *Ash dieback*

Comments

1. *Fraxinus excelsior* suffers from a condition, not fully explained, called Ash dieback. It has not been clearly characterized but involves the death of scattered twigs, branches or limbs (Fig.7). Even very severely affected trees sometimes slowly recover. It is common in hedgerows, particularly those adjacent to arable land or ditches; its incidence is higher in the eastern, drier parts of the country. Ash bud moth (p.68) is frequently present in affected trees and may be involved in the syndrome to some degree but the prime suspect is root damage caused by ploughing. If a thorough and systematic investigation into a case of dieback fails to reveal another cause, the possibility that this 'Ash dieback' is the problem should be considered.

2. Small knobbly growths found sometimes on the smaller branches are known as 'Ash roses'. These occur where the Ash bark beetle (*Hylesinus varius*) tunnels into the bark and wood to overwinter.

3. Verticillium wilt (p. 261, caused by *V. dahliae*), but without the usual characteristic wood staining, is said to be widespread in Holland on *Fraxinus excelsior* and its cultivars of all ages. In the USA, the same fungus causes leaf scorch and premature defoliation of *F. americana*, *F. pennsylvanica* and *F. oxycarpa* 'Raywood', again in the absence of wood staining. The disease has not been confirmed on ash in this country, however.

GINKGO BILOBA – Maidenhair Tree

■ **Roots, root collar, butt**

Honey fungus killing	145
Root and butt rots: *Ganoderma adspersum/applanatum, Meripilus giganteus, Armillaria*	226

Comments

1. In the USA, Ginkgo is regarded as largely disease and pest free but we have seen apparently primary attacks of both Honey fungus and *M. giganteus* on sizeable trees in this country.

2. In the USA, in one particular incident gingko proved to be very sensitive to flooding by sea water.

[Roots: *Heterobasidion annosum* not recorded – see p.220]

GLEDITSIA – Honey Locust

■ **Roots, root collar, butt**

Root rot: *Meripilus giganteus*   229

■ **Foliage (leaves, shoots, buds), flowers, fruit**

Leaf galls   132,149

■ **Structural and water-conducting wood**

Watery exudate from pruning and other wounds   103

Comments

In its native USA, *Gleditsia triacanthos* is rarely troubled by serious disease.

[Roots: *Heterobasidion annosum* not recorded – see p.220].

GRISELINIA – Griselinia

■ **Bark, cambium of stem, branches, twigs**

Winter cold damage   129

■ **Foliage (leaves, shoots, buds), flowers, fruit**

Winter cold damage   129

Comments

[Roots: *Heterobasidion annosum* not recorded – see p.220]

GYMNOCLADUS – Kentucky Coffee Tree

■ **Structural and water-conducting wood**

Watery exudate from pruning and other wounds   103

Comments

Gymnocladus dioicus is regarded in its native USA as remarkably disease free.

[Roots: *Heterobasidion annosum* not recorded – see p.220].

HALESIA – Snowdrop Trees

■ **Foliage (leaves, shoots, buds), flowers, fruit**

Lime induced chlorosis   174

Comments

[Roots: *Heterobasidion annosum* not recorded – see p.220]

ILEX – Hollies

■ **Roots, root collar, butt**

See Comment (1) below

■ **Bark, cambium of stem, branches, twigs**

Perennial cankers   186

■ **Foliage (leaves, shoots, buds), flowers, fruit**

Phytophthora blight   203
Leaf mines   143

See also Comment (2) below

Comments

1. There are so few reported cases of Honey fungus (*Armillaria*) killing of

the very common holly, *Ilex aquifolium*, that it can be regarded as very resistant to the disease.

2. The reason for the quite common fall of large numbers of green, apparently healthy leaves from *I. aquifolium* is not known.

[Roots: *Heterobasidion annosum* not recorded – see p.220]

JUGLANS – Walnuts

■ **Roots, root collar, butt**

Honey fungus killing (see also Comment below)	145
Phytophthora killing	206

[*Heterobasidion annosum* not recorded – see p.220]

■ **Bark, cambium of stem, branches, twigs**

Coral spot fungus	106
Black line disease (in grafted trees)	82

■ **Foliage (leaves, shoots, buds), flowers, fruit**

Leaf spots	164,165
White leaf spot	165
Leaf galls	133
Late (spring) frost damage	126
Honeydew and Sooty moulds	144

■ **Structural and water-conducting wood**

Watery exudate from pruning and other wounds	103

Comments

1. English or Persian walnut, *Juglans regia*, is particularly susceptible to killing by Honey fungus (*Armillaria*), whereas observations and experimentation indicate that *J. hindsii* and *J. nigra* are probably immune.

2. See allelopathic effects of *Juglans* under Harmful Chemicals on p. 6

[Roots: *Heterobasidion annosum* not recorded – see p.220]

JUNIPERUS – Junipers

■ **Roots, root collar, butt**

Phytophthora root killing	206

■ **Bark, cambium of stem, branches, twigs**

Phomopsis canker	201
Stem rust disease	244
See also Amylostereum 'status'	64

■ **Foliage (leaves, shoots, buds), flowers, fruit**

Kabatina shoot blight	152
Phomopsis canker	201
Pestalotia foliage blight	153
Scale insects	249
Mined shoots	151
Mined needles – see Comment (1) below	
Honeydew and Sooty moulds	144
– see Comments (2) and (3) below	
Urine damage	93
Foliage bronzing	130
See also Comment (4) below and Amylostereum 'status'	64

Comments

1. Young larvae of the Juniper webber moth, *Dichomeris marginella*, mine the needles in September. This damage is inconspicuous but, after overwintering, the older and by now larger, striped larvae (10mm long), feeding on the needles in May and June from within the protection of a dense silk web, cause much browning of the foliage and some shoot dieback (Fig.8).

2. Heavy infestations of the American juniper aphid, *Cinara fresai*, on stems in the centre of small *J. chinensis*, *J. sabina*, *J. squamata*, and *J.*

2

SECTION

Fig. 8 *Juniper needles damaged by larvae of the moth Dichomeris marginella*

virginiana kill the internal foliage, leading to dieback and sometimes death of the whole tree. In these cases the aphids are usually plain to see.

3. The Juniper aphid, *Cinara juniperi*, occurs on the current shoots of *J. communis* and other species but does not cause dieback.

4. The shiny, raised, black spots common on dead scale leaves of junipers are the fruit bodies of the wholly saprophytic fungus, *Lophodermium juniperi*.

KALOPANAX – Prickly Castor-oil Tree

■ **Foliage (leaves, shoots, buds), flowers, fruit**

Early (autumn) frost damage   129

■ **Structural and water-conducting wood**

Verticillium wilt   261

Comments

[Roots: *Heterobasidion annosum* not recorded – see p.220]

KOELREUTERIA – Pride of India

■ **Bark, cambium of stem, branches, twigs**

Coral spot fungus   106

■ **Structural and water-conducting wood**

Verticillium wilt   261

Comments

[Roots: *Heterobasidion annosum* not recorded – see p.220]

+LABURNOCYTISUS 'ADAMII' Adam's Laburnum

Comments

A lax witch's broom closely resembling those caused by *Taphrina wiesneri* on cherry (Figs. 399 and 401) has been noted once on this graft hybrid tree.

The authors have traced no other records of pests, disease or disorders on it, but it would not be surprising if it were liable to those which affect *Laburnum*, one of its parents.

[Roots: *Heterobasidion annosum* not recorded – see p.220]

LABURNUM – Laburnum

■ **Roots, root collar, butt**

Phytophthora killing   206

See also Comments (1) and (2) below

■ **Bark and cambium of stem, branches, twigs**

Silver leaf killing   252

■ **Foliage (leaves, shoots, buds), flowers, fruit**

Powdery mildew   215
Leaf mines   159

For problems affecting many tree genera, see p.24

See also Silver leaf   252

- **Structural and water-conducting wood**

  Silver leaf wood staining   252
  Watery exudate from pruning or
  other wounds   103

Comments

1. Several cases have been noted where Laburnum has failed to flush in spring yet the buds and bark have remained green for the whole summer. The roots and stem base below ground were found to have been killed by *Phytophthora* (p.206).

2. There are very few records of Honey fungus killing *Laburnum*.

[Roots: *Heterobasidion annosum* not recorded – see p.220]

LARIX – Larches

- **Roots, root collar, butt**

  Honey fungus killing   145
  Root rots: *Heterobasidion annosum,
  Hypholoma fasciculare, Armillaria*
  Butt rots: *Phaeolus schweinitzii,
  Sparassis crispa, Armillaria*   226

- **Bark and cambium of stem, branches, twigs**

  Phacidium twig killing. See
  Comment (2) below
  Larch dieback   155

  See also Comment (1) below

- **Foliage (leaves, shoots, buds), flowers, fruit**

  Needle cast disease   187
  Woolly aphids   155
  Leaf mines   158
  Drought damage   112
  Lime induced chlorosis   174
  Honeydew and Sooty moulds   155

  See also Comment (2) below

- **Structural and water-conducting wood**

  Stem rots: *Phaeolus schweinitzii,
  Sparassis crispa*   226

Fig. 9 *Fruit bodies of the Larch canker fungus Lachnellula willkommii*

Comments

1. Certain (High Alpine) provenances of *Larix decidua* are liable to cankering by the fungus *Lachnellula willkommii*. Perennial 'target' cankers, eventually very large, form on branches and main stems. In appearance and effect the disease resembles Nectria canker (p.179) though the conspicuous fruit bodies are like tiny jam tarts with an orange centre and white rim (Fig.9).

Fig. 10 *Larch shoots killed by a species of the fungus Cytospora*

2. In spring and summer, brown needles may draw attention to scattered dead extension shoots, usually one, sometimes two years old. If the deaths are due to girdling patches of dead bark at the base of the shoots, a

2

SECTION

For problems affecting many tree genera, see p.24

species of the fungus *Cytospora* (= *Leucocytospora*) (probably *C. kunzei*) is the likely cause (Fig.10). No control is required. Similar damage is sometimes caused by *Phacidium coniferarum* but this also attacks older wood (see under *Cedrus* on p.30, and Fig.4). Current year's shoots are sometimes mined by larvae of the moth *Argyresthia laevigatella*.

LAURUS – Bay Laurel or Sweet Bay

■ **Bark, cambium of stem, branches, twigs**

Scale insects	149

■ **Foliage (leaves, shoots, buds), flowers, fruit**

Winter cold damage	129
Leaf galls	133

See also Comments below

Fig. 11 *Shoot of Bay laurel girdled by the fungus Diaporthe nobilis*

Comments

Occasionally in summer, the leaves of a number of one-year-old shoots of *Laurus nobilis* wither and hang brown and dead on the tree. The cause is the fungus *Diaporthe nobilis* (anamorph: *Phomopsis laurella*) which kills a ring of bark part way along or at the base of the shoot (Fig.11). The disease does not seem to have been investigated further. No control is required.

LIGUSTRUM LUCIDUM – Glossy or Chinese Privet

■ **Roots, root collar, butt**

White Root rot	146

See Comments below

■ **Foliage (leaves, shoots, buds), flowers, fruit**

See Comments below

Comments

Information is lacking on the pathology of the only tree-sized species considered here – *Ligustrum lucidum*. Whether it is liable to the same problems as *L. vulgare* and *L. ovalifolium* is not known. These common hedging shrub species are very susceptible to Honey fungus killing (p.145) and *L. vulgare* is susceptible to killing by the root rotting fungus, *Heterobasidion annosum* (p.220); to damage by a leaf miner (p.158); and to the Privet thrips, *Dendrothrips ornatus* a sucking insect which causes a silvering or silvery stippling on the leaves. (cf Leafhopper and Mite damage on p.157).

LIQUIDAMBAR – Sweet Gums

■ **Roots, root collar, butt**

Root and butt rot: *Ustulina deusta*	208

■ **Bark, cambium of stem, branches, twigs**

Corky wings	103

■ **Foliage (leaves, shoots, buds), flowers, fruit**

Lime induced chlorosis	174
Salt damage	93

For problems affecting many tree genera, see p.24

LIRIODENDRON – Tulip Trees

■ **Roots, root collar, butt**

Honey fungus killing	145
Waterlogging	266

See Comments below

■ **Foliage (leaves, shoots, buds), flowers, fruit**

Leaf spots	114
Salt damage	93

■ **Structural and water-conducting wood**

Bacterial wetwood	74

Comments

In its native USA, *Liriodendron tulipifera* is known to be particularly intolerant of flooding and of changes in soil level.

[Roots: *Heterobasidion annosum* not recorded – see p.220]

MACLURA – Osage Orange

■ **Structural and water-conducting wood**

Verticillium wilt	261

Comments

[Roots: *Heterobasidion annosum* not recorded – see p.220]

MAGNOLIA – Magnolias

■ **Roots, root collar, butt**

Waterlogging	266

■ **Bark, cambium of stem, branches, twigs**

Coral spot fungus	106
Scale insects	149,250

■ **Foliage (leaves, shoots, buds), flowers, fruit**

Leaf spots	161
Spring frost damage (flowers)	126
Lime induced chlorosis	174

■ **Structural and water-conducting wood**

Watery exudate from pruning or other wounds	103

See also Comments below

Comments

Verticillium wilt (p.261) has been recorded on Magnolia in the USA where it is reported that the stain in the wood, which is so characteristic of the disease, was absent.

[Roots: *Heterobasidion annosum* not recorded – see p.220]

MALUS – Apples

See Comment (1) below

■ **Roots, root collar, butt**

Crown Gall	135
Honey fungus killing	145
Root and butt rots: *Ganoderma* species, *Armillaria*	226
White root rot	146

■ **Bark, cambium of stem, branches, twigs**

Crown Gall	135
Nectria canker	185
Blossom wilt, spur blight and wither tip	85
Silver leaf killing	252
Fireblight killing	124
Scale insects	250
Woolly aphids	276
Cherry bark tortrix damage	98
Papery bark – see Comment (3) below	

■ **Foliage (leaves, shoots, buds), flowers, fruit**

Scab/Leaf spots	245

SECTION 2

■ **Structural and water-conducting wood**

Fig. 12 *Papery bark disease of apple*

Comments

1. Diagnostic difficulties are quite often experienced as a result of mistaking Flowering crabs (*Malus*) for Flowering cherries (*Prunus*).

2. Orchard apples are liable to Phytophthora root disease (p.206) but this disease has not come to our notice on Flowering crab apples.

3. Thin, papery, outer layers of bark on twigs and branches may peel away (Fig.12). The underlying bark is dead. This may be associated with Silver leaf (p.252) but in budded *Malus*, especially *M. tschonoskii* on

rootstock M.9, this is likely to be the physiological 'Papery bark' disorder. It is suggested that pruning removes the parts into which growth hormones would normally pass in spring. Where these accumulate behind the pruning wounds, abnormal callus tissue forms then dies. Damage follows winter or early spring pruning but is slight following pruning in early winter or May.

MESPILUS – Medlar

■ **Foliage (leaves, shoots, buds), flowers, fruit**

Comments

[Roots: *Heterobasidion annosum* not recorded – see p.214]

METASEQUOIA GLYPTOSTROBOIDES – Dawn Redwood

■ **Roots, root collar, butt**

■ **Foliage (leaves, shoots, buds), flowers, fruit**

MORUS – Mulberries

■ **Roots, root collar, butt**

■ **Bark, cambium of stem, branches, twigs**

For problems affecting many tree genera, see p.24

See also Comments below

- **Foliage (leaves, shoots, buds), flowers, fruit**

  Salt damage — 93

See also Comments below

Comments

A bacterial blight of mulberry occurs in Britain but has not been recorded for many years. Leaves develop black spots surrounded by a yellow border; shoots develop black lesions from which, in wet weather, bacterial slime oozes. Severe attacks cause a general dieback. For further information see Peace (1962) and Smith et al (1988) under *Pseudomonas syringae* pv. *mori*.

[Roots: *Heterobasidion annosum* not recorded – see p.220]

MYRTUS – Myrtles

- **Bark, cambium of stem, branches, twigs**

  Winter cold damage — 129

- **Foliage (leaves, shoots, buds), flowers, fruit**

  Leaf spots — 161
  Winter cold damage — 129

Comments

[Roots: *Heterobasidion annosum* not recorded – see p.220]

NOTHOFAGUS – Southern Beech

- **Roots, root collar, butt**

  Root rot: *Heterobasidion annosum* — 220
  Phytophthora killing — 206

- **Bark, cambium of stem, branches, twigs**

Winter cold damage — 129

- **Foliage (leaves, shoots, buds), flowers, fruit**

  Lime induced chlorosis — 174
  Leafhopper damage — 157

NYSSA – Tupelos

- **Bark, cambium of stem, branches, twigs**

  Stereum canker-rot — 259

- **Foliage (leaves, shoots, buds), flowers, fruit**

  Lime induced chlorosis — 174
  Salt damage — 93

OSTRYA – Hop Hornbeams

- **Roots, root collar, butt**

  Waterlogging — 266

- **Foliage (leaves, shoots, buds), flowers, fruit**

  Salt damage — 93

Comments

In the USA, where it is indigenous, *Ostrya virginiana* is host to the leaf spot fungus, *Gnomoniella carpinea*, which in this country occurs on the related *Carpinus betulus* (see p.164).

[Roots: *Heterobasidion annosum* not recorded – see p.220]

OXYDENDRUM – Sorrel Tree

- **Foliage (leaves, shoots, buds), flowers**

  Lime induced chlorosis — 174

Comments

[Roots: *Heterobasidion annosum* not recorded – see p.220]

SECTION **2**

PARROTIA – Persian Ironwood

Comments

[Roots: *Heterobasidion annosum* not recorded – see p.220]. The authors have found no other useful information on the pathology of Parrotia

PAULOWNIA TOMENTOSA – Foxglove tree

■ **Roots, root collar, butt**

Honey fungus	145
Phytophthora root disease	206
See also Comment below	

■ **Foliage (leaves, shoots, buds), flowers, fruit**

Spring frost (flowers)	126

■ **Structural and water-conducting wood**

Wind breakage of branches	176

Comments

[Roots: *Heterobasidion annosum* not recorded - see p. 220]

PHELLODENDRON – Phellodendron

■ **Foliage (leaves, shoots, buds), flowers, fruit**

Spring frost	126

Comments

[Roots: *Heterobasidion annosum* not recorded - see p. 220]

PHILLYREA – Phillyrea

■ **Foliage (leaves, shoots, buds), flowers, fruit**

Leaf mines – see Comments below

Comments

Phillyrea leaves are sometimes mined by the Lilac leaf miner, *Caloptilia syringella* (see p158 for a general account of leaf mines).

[Roots: *Heterobasidion annosum* not recorded – see p.220]

PICEA – Spruces

■ **Roots, root collar, butt**

Honey fungus killing	145
Root and butt rots: *Heterobasidion annosum, Phaeolus schweinitzii, Sparassis crispa, Hypholoma fasciculare, Armillaria*	226
Bark beetle damage	138
Woolly aphids (roots)	273

■ **Bark, cambium of stem, branches, twigs**

Bark beetle damage	138
see also Eight-toothed spruce bark beetle	281
See also Comments below	

■ **Foliage (leaves, shoots, buds, flowers), fruit**

Chrysomyxa leaf rust fungi	188
Aphid damage (needles)	140
Mite damage (needles)	105,119
Needle mines	160
Adelgid galls	135
Woolly aphids	273
Honeydew and Sooty moulds	140
Late (spring) frost damage	126
Lime induced chlorosis (some spp.)	174
Cucurbitaria bud blight	87
see also White pine weevil	282

See also Comments below

■ **Structural and water-conducting wood**

Stem rots: *Phaeolus schweinitzii, Sparassis crispa, Stereum sanguinolentum*	226

Fig. 13 *Unexplained branch death of Blue spruce*

Comments

1. *Picea abies* planted as an isolated specimen tree or in other exposed situations often deteriorates after a number of years of normal growth: shoot growth becomes very poor, giving the top of the crown a con-certinaed look after a while, needles brown, shoots die, and sometimes the whole tree dies from the top downwards. This is the condition known as Top dying of Norway spruce. The precise cause is unclear but growth depression sets in following an unusually mild winter while the dieback has to do with exposure and disturbed water relations. There are no control measures. See Gregory and Redfern 1998.

2. The death of main branches on otherwise healthy *Picea pungens glauca* (Fig.13), sometimes attributed to Green spruce aphid infestations, has not been satisfactorily explained. The possibility that the needle fungus, *Rhizosphaera kalkhoffii*, is involved requires investigation.

3. Extensive colonies of *Cinara piceae*, a large, tarry black aphid, sometimes occur on the stem and larger branches of *Picea abies* and other spruces. Apart from their nuisance value they are no cause for concern.

PICRASMA – Picrasma

Comments

[Roots: *Heterobasidion annosum* not recorded – see p.220]. The authors have found no other useful information on the pathology of *Picrasma*.

PINUS – Pines

Roots, root collar, butt

Honey fungus killing	145
Woolly aphids (roots)	273
Root and butt rots: *Heterobasidion annosum, Phaeolus schweinitzii, Sparassis crispa*	226

Bark, cambium of stem, branches, twigs

Stem rusts	83, 84
Pine shoot beetle damage	209
Woolly aphids	273

Foliage (leaves, shoots, buds), flowers, fruit

Needle cast diseases	189,190
Pine shoot beetle damage	209
Pine shoot moth damage	211
Honeydew and Sooty moulds	144
Lime induced chlorosis	174
see also White pine weevil	282

See also Comments (1) and (2) below

Structural and water-conducting wood

Stem rots: *Heterobasidion annosum, Phaeolus schweinitzii, Sparassis crispa, Phellinus pini*	226

SECTION **2**

For problems affecting many tree genera, see p.24

Comments

1. Scattered one-year-old shoots of 2-needled pines, notably *Pinus nigra var. nigra*, are occasionally killed by the fungus *Sphaeropsis sapinea*. Abroad this is sometimes a serious disease but here damage is slight and no control is required.

2. Scattered shoot and branch dieback of *Pinus nigra var. maritima* and *P. sylvestris*, caused by the fungus *Gremmeniella* (*Brunchorstia*), quite common in forest plantations, may occasionally be encountered in amenity trees. See Redfern and Gregory (1998).

3. In *Pinus sylvestris* during summer, the fall of scattered pairs of current needles, characteristically splayed and often caught up in the remaining foliage, is due to feeding damage within the needle sheath by the Pine needle midge, *Contarinia baeri*. No control is required.

PITTOSPORUM – Pittosporum

■ **Bark, cambium of stem, branches, twigs**

Winter cold damage — 129

■ **Foliage (leaves, shoots, buds), flowers, fruit**

Winter cold damage — 129
Late (spring) frost damage — 126

Comments

Discoloration and blistering of the young foliage accompanied by heavy deposits of honey dew and sooty moulds (see p. 144) indicate feeding by the New Zealand Pittosporum psyllid, *Trioza vitreoradiata*. The green, elliptical, flattened nymphs fit neatly into the concave undersides of the blisters that their feeding causes. The 3-4 mm long, winged adults are yellowish green to black. First noticed here in 1993, it is now established in Cornwall and the Isles of Scilly and could appear wherever Pittosporum is grown. Although its effects are an unsightly nuisance, the insect requires control (with systemic insecticides) only if the foliage is being grown for sale.

[Roots: *Heterobasidion annosum* not recorded – see p.220]

PLATANUS – Plane Trees

■ **Roots, root collar, butt**

See Comments below

■ **Bark, cambium of stem, branches, twigs**

Anthracnose — 66
Salt damage — 93
see also Canker stain of plane — 280

■ **Foliage (leaves, shoots, buds), flowers, fruit**

Anthracnose — 66
Powdery mildew — 215
Leaf mines — 160
Salt damage — 93

■ **Structural and water-conducting wood**

Stem rots: *Inonotus hispidus*,
Ganoderma species — 226
see also Canker stain of plane — 280

Comments

The exceedingly few cases of Honey fungus that have been reported on this commonly planted amenity tree genus, both in this country and in the USA, indicate that it is one of the more resistant genera to the disease. One case of *Armillaria mellea* killing of a number of large London plane in England has been noted, however.

[Roots: *Heterobasidion annosum* not recorded – see p.220]

PODOCARPUS – Yellow-woods

■ **Bark, cambium of stem, branches, twigs**

Winter cold damage   129

■ **Foliage (leaves, shoots, buds), flowers, fruit**

Winter cold damage   129

Comments

[Roots: *Heterobasidion annosum* not recorded – see p.220]

POPULUS – Poplars

■ **Roots, root collar, butt**

White root rot   146

■ **Bark, cambium of stem, branches, twigs**

Bacterial canker	72
Longhorn beetle damage	255
Cytospora spore tendrils	102
Squirrel damage	178

See also Comment (1) below re
P. nigra 'Italica' and Comment (2)

■ **Foliage (leaves, shoots, buds), flowers, fruit**

Leaf spots	165
Scab	245
Leaf blisters	197
Leaf rust fungi	241
Leaf beetle damage	156
Leaf mines	158
Leaf galls	133,134
Petiole galls	134
Shoot borer damage	214
Lime induced chlorosis	174

■ **Structural and water-conducting wood**

Bacterial wetwood	74
Watery exudate from pruning or other wounds	103

Top rot: *Chondrostereum purpureum*	226,254
Wind snap	176
Twig shedding – see Comment (3) below	
Mistletoe	103

Fig. 14 *Unexplained dieback of Lombardy poplar*

Comments

1. The not uncommon death of branches of *P. nigra* 'Italica', particularly in the upper crown, has not been explained (Fig.14). It has the hallmarks of the disease attributed in the USA and on the Continent to the bark-killing fungus, *Cryptodiaporthe populea* (anamorph: *Discosporium populea*, syn. *Dothichiza populea*). This does occur in Britain but has not been found associated with the disease in question. No control is available.

2. Holes, circular, ragged and about 8mm in diameter, occur at the trunk base where adults of the Hornet moth, *Sesia apiformis*, have emerged. The pale yellow or creamy larvae have brown heads, are up to 40 mm long and feed in the bark cambium

SECTION 2

around the root collar. A link with dieback has been suggested but is not proven.

3. In summer, numerous fallen twigs, often with attached, dead leaves, sometimes litter the ground beneath apparently healthy trees of *Populus* and other broadleaved genera. The twig has usually become detached at its base and the break is smooth, dome-shaped and shows no sign of chewing or other mechanical injury. This is a natural shedding, termed cladoptosis, of moribund, shaded twigs. Drought may accelerate the process.

PRUNUS – Cherries, Plums, Laurels etc.

See Comment (1) below

■ Roots, root collar, butt

Crown Gall	135
Honey fungus	145
Phytophthora root disease	206
White root rot	146
Waterlogging	266

■ Bark, cambium of stem, branches, twigs

Bacterial canker	70
Crown Gall	135
Silver leaf bark killing	252
Dieback of Prunus 'Kanzan'	216
Cytospora dieback of Cherry laurel	110
Cherry bark tortrix damage	98
Scale insects	149,250
Death of previous year's twigs	85

■ Foliage (leaves, shoots, buds), flowers, fruit

Blossom wilt, spur blight and wither tip	85
Leaf thickening and distortion	197,270
Leaf blotches or spots	166
Defoliation	166
Leaves retained in winter	166

Distorted fruits	213
Powdery mildew	215
Shothole in leaves	71,110,166
Blackfly damage	99
Leaf mines	158
Leaf galls	133
Leafhopper damage	157
Slugworm damage	198
Bird damage to buds	178
Lime induced chlorosis	174
Honeydew and Sooty moulds	144

See also Comment (2) below.

■ Structural and water-conducting wood

Silver leaf wood staining	252
Top rots: *Chondrostereum purpureum, Coriolus versicolor, Ganoderma* species, *Laetiporus sulphureus, Phellinus tuberculosus*	226

■ Other

Witches' brooms	270

Comments

1. Difficulties are quite often experienced as a result of mistaking Flowering crabs (*Malus*) for Flowering cherries (*Prunus*).

2. The upper surfaces of plum and blackthorn leaves are sometimes covered in bright yellow spots. Heavily infected leaves fall prematurely. The cause is the rust fungus, *Tranzschelia pruni-spinosae*. No control is necessary.

PSEUDOLARIX – Golden Larch

■ Bark, cambium of stem, branches, twigs

Larch canker	41

■ Foliage (leaves, shoots, buds), flowers, fruit

Lime induced chlorosis	174

For problems affecting many tree genera, see p.24

Comments

[Roots: *Heterobasidion annosum* not recorded – see p.220]

PSEUDOTSUGA – Douglas Firs

- **Roots, root collar, butt**

  Butt rots: *Phaeolus schweinitzii,*
  Sparassis crispa   226

- **Bark, cambium of stem,**
  branches, twigs

  Phacidium branch killing. See
  Cedrus, Comment (1).
  See also Comments below.

- **Foliage (leaves, shoots, buds),**
  flowers, fruit

Needle cast diseases	191
Woolly aphids	273
Winter cold damage	129
Lime induced chlorosis	174
Salt damage	93

- **Structural wood**

  Stem rots: *Phaeolus schweinitzii,*
  Sparassis crispa   226

Comments

An unexplained disease of Douglas fir characterized by severe resin bleeding from the stem, slow crown deterioration and sometimes death of the tree occurs occasionally. In Germany, similar symptoms have been linked to an excess of manganese.

PTEROCARYA – Wing-nuts

- **Foliage (leaves, shoots, buds),**
  flowers, fruit

Late (spring) frost damage	126

- **Structural and water-conducting**
  wood

Watery exudate from pruning or
other wounds   103

Comments

[Roots: *Heterobasidion annosum* not recorded – see p.220]

PYRUS – Pears

- **Roots, root collar, butt**

White root rot	146
Crown gall	135

- **Bark, cambium of stem,**
  branches, twigs

Crown Gall	135
Fireblight bark killing	124
Silver leaf bark killing	252
Nectria canker	185
Blossom wilt, spur blight and wither tip	85
Cherry bark tortrix damage	98
Woolly aphids	275

- **Foliage (leaves, shoots, buds),**
  flowers, fruit

Scab	245
Powdery mildew	215
Fireblight killing	124
Blossom wilt and wither tip	85
Leaf blister mite damage	200
Slugworm damage	198
Leaf galls	133

- **Structural and water conducting**
  wood

Silver leaf wood staining	252
Top rots: *Chondrostereum purpureum*	226,252

Comments

Leaves, and occasionally fruits, of *Pyrus* are sometimes bronzed by the Rust mite, *Epitrimerus piri*. No control is required.

SECTION 2

For problems affecting many tree genera, see p.24

QUERCUS – Oaks

■ **Roots, root collar, butt**

Root and butt rots: *Collybia fusipes, Fistulina hepatica, Ganoderma resinaceum, G. adspersum, G. applanatum, Grifola frondosa, Inonotus dryadeus, Hypholoma fasciculare*   226

See also Comment (2) below

■ **Bark, cambium of stem, branches, twigs**

Stereum canker-rot	259
Bulgaria canker	91
Scale insects	251
Oak dieback	192

■ **Foliage (leaves, shoots, buds), flowers, fruit**

Oak dieback	192
Oak mildew	215
Leaf spots and blotches	167
Oak leaf phylloxera damage	194
Galls on leaves, buds, flowers	133,134,135
Galled acorns	153
Leafhopper damage	157
Leaf mines	159
Honeydew and Sooty moulds	144
Late (spring) frost damage	126
Drought damage	112
Lime induced chlorosis	174

See also Comments (1) and (3) below

■ **Structural and water-conducting wood**

Top rots: *Laetiporus sulphureus, Ganoderma adspersum, G. applanatum*   226
Leopard moth damage (twigs and small branches)   169
Twig shedding – see Comment (3) under *Populus*
see also Oak wilt   281

Fig. 15 *Natural retention of dead leaves in winter by English oak*

Comments

1. Slugworm larvae of *Caliroa cinxia* partially remove the epidermis of leaves which then appear translucent and dirty white in colour (see also p.196).

2. The oaks are among the more Honey fungus resistant trees.

3. Young deciduous oaks often retain many of their dead leaves over winter (Fig.15). This is not a pathological condition.

RHUS VERNICIFLUA – Varnish Tree

■ **Foliage (leaves, shoots, buds), flowers, fruit**

Late (spring) frost damage   126

■ **Structural and water-conducting wood**

See Comments below

Comments

There appears to be virtually no

For problems affecting many tree genera, see p.24

published information on the pathology of the only species considered here, *Rhus verniciflua*. Whether this species is susceptible to Verticillium wilt (p.241), as *R. typhina* is, is not known.

ROBINIA – Locust Trees

■ **Roots, root collar, butt**

Phytophthora killing	206
Root and butt rots: *Laetiporus* *sulphureus, Perenniporia fraxinea,* *Pholiota squarrosa*	226

■ **Bark, cambium of stem, branches, twigs**

Scale insects	250

■ **Foliage (leaves, shoots, buds), flowers, fruit**

Rust mites – see Comments below

■ **Structural and water-conducting wood**

Watery exudate from pruning or other wounds	103
Wind breakage of branches	176

Comments

Leaves of *Robinia pseudoacacia* are sometimes discoloured and distorted by False acacia rust mites, *Vasates albotrichus* and *V. robiniae* (Alford 1991). No control is required.

[Roots: *Heterobasidion annosum* not recorded – see p.220].

SALIX – Willows

■ **Roots, root collar, butt**

Honey fungus killing and rot	145

■ **Bark, cambium of stem, branches, twigs**

Scab	245
Black canker (*Glomerella*)	245

Anthracnose	167
Scale insects	149,250
Longhorn beetle damage	235

■ **Foliage (leaves, shoots, buds), flowers, fruit**

Scab	245
Anthracnose	167
Black canker (*Glomerella*)	245
Rust fungi	241
Leaf beetle damage	156
Leaf galls	133
Leaf mines	158
Honeydew and Sooty moulds	144
Catkin galls (i.e. small Witches' brooms)	184

■ **Structural and water-conducting wood**

Watermark disease	267
Top rot: *Daedaleopsis confragosa*	226
Twig shedding – see Comment (3) under *Populus*	

SASSAFRAS – Sassafras

■ **Foliage (leaves, shoots, buds), flowers, fruit**

Late (spring) frost damage	126
Lime induced chlorosis	174
Salt damage	93

Comments

[Roots: *Heterobasidion annosum* not recorded – see p.220]

SAXEGOTHAEA – Prince Albert's Yew

Comments

[Roots: *Heterobasidion annosum* not recorded – see p.220]. The authors have found no other useful information on the pathology of *Saxogothaea*.

SECTION 2

SCIADOPITYS –
Japanese Umbrella Pine

■ **Foliage (leaves, shoots, buds), flowers, fruit**

Lime induced chlorosis   174

Comments

[Roots: *Heterobasidion annosum* not recorded – see p.220]

SEQUOIA – Coast Redwood

■ **Foliage (leaves, shoots, buds), flowers, fruit**

Winter cold damage   129

Comments

The Coast redwood is said in the USA to have had fewer foliar pathogens reported on it than any other major tree species in that country.

SEQUOIADENDRON – Wellingtonia

■ **Roots, root collar, butt**

Honey fungus killing and rot   145

■ **Bark, cambium of stem, branches, twigs**

Lightning   170

■ **Foliage (leaves, shoots, buds), flowers, fruit**

Scale insects   249

SOPHORA – Pagoda Tree

■ **Bark, cambium of stem, branches, twigs**

Winter cold damage   129

■ **Foliage (leaves, shoots, buds), flowers, fruit**

Winter cold damage   129
Late (spring) frost damage   126

■ **Structural and water-conducting wood**

Watery exudate from pruning or other wounds   103

Comments

[Roots: *Heterobasidion annosum* not recorded – see p.220]

SORBUS (Aria Section) – Whitebeams

■ **Roots, root collar, butt**

Root and butt rot: *Ganoderma adspersum, G. applanatum*   226

■ **Bark, cambium of stem, branches, twigs**

Fireblight killing   124
Woolly aphids   275
Graft incompatibility   177
Stereum canker-rot   259

■ **Foliage (leaves, shoots, buds), flowers, fruit**

Fireblight killing   124
Scab   245
Leaf galls   200
Leaf mines   158
Drought damage   112

■ **Structural and water-conducting wood**

Graft incompatibility   177

Comments

Forestry Commission records indicate that Whitebeam (*S. aria*) is much more resistant to Honey fungus than Rowan (*S. aucuparia*).

SORBUS (Aucuparia Section) – Rowans

■ **Roots, root collar, butt**

Honey fungus killing   145

■ **Bark, cambium, branches, twigs**

Nectria canker	185
Fireblight killing	124
Silver leaf bark killing	252
Woolly aphids	272

Death of twigs – see Comment (1) below

■ **Foliage (leaves, shoots, buds), flowers, fruit**

Fireblight killing	124
Leaf rust fungi	244
Scab	245
Leaf galls	134,200
Leaf mines	158
Slugworm damage	199
Lime induced chlorosis	174

Leaf discoloration, leafdrop – see Comment (2) below

■ **Structural and water-conducting wood**

Silver leaf staining	252
Top rot: *Chondrostereum purpureum*	226

Fig. 16 *Wild service tree leaves damaged by aphids.*

Comments

1. A canker and dieback of exotic rowan species has been reported from Scotland, apparently caused by the fungus *Valsa cincta* (anamorph: *Cytospora rubescens*), a pathogen of some importance on environmentally stressed peach, apricot and other *Prunus* species abroad.

2. In May/June scattered leaves or patches of leaves on *S. torminalis* infested by the aphid *Dysaphis aucupariae* are pale yellow-green, later becoming tinged with red before turning golden brown (see Fig. 16). Severely damaged leaves are rolled or twisted. Some twig dieback can result from such attacks. The aphids soon disappear from the discoloured leaves.

STUARTIA – Stuartia or Stewartia

■ **Foliage (leaves, shoots, buds), flowers, fruit**

Lime induced chlorosis	174

Comments

[Roots: *Heterobasidion annosum* not recorded – see p.220]

STYRAX – Storax

■ **Foliage (leaves, shoots, buds), flowers, fruit**

Late (spring) frost damage	126
Lime induced chlorosis	174

Comments

[Roots: *Heterobasidion annosum* not recorded – see p.220]

TAXODIUM – Deciduous Cypresses

■ **Foliage (leaves, shoots, buds), flowers, fruit**

Lime induced chlorosis	174

Comments

[Roots: *Heterobasidion annosum* not recorded – see p.220]

SECTION 2

TAXUS – Yews

■ **Roots, root collar, butt**

Phytophthora killing 206
Root and butt rot: *Ganoderma valesciacum sensu* British mycologists 222

See Comment (1) below

■ **Bark, cambium of stem, branches, twigs**

Dieback of odd branches 277
See also Amylostereum 'status' 64

■ **Foliage (leaves, shoots, buds), flowers, fruit**

Scale insects 250,251
Artichoke galls 279
Ditula shoot killing 131
Honeydew and Sooty moulds 144
Drought damage 112
Hormone herbicide damage 93
Autumn frost bronzing 130

See also Comment (2) below

■ **Structural and water-conducting wood**

Top rot: *Laetiporus sulphureus* 226

Comments

1. Authenticated cases of Honey fungus killing of *Taxus* are extremely rare.

2. Small, longitudinally-split lumps on the underside of green or yellowed needles of *T. baccata* resemble but are not fungal fruit bodies. Their cause and significance are not known, but evidence has been put forward that similar structures are caused variously by an excess of water in the tissues ('oedema'), by mite feeding, or by fungal infection.

TETRACENTRON – Spur Leaf

Comments

[Roots: *Heterobasidion annosum* not recorded – see p.220]. The authors have found no other useful information on the pathology of *Tetracentron*.

TETRADIUM (=EUODIA) – Euodia or Evodia

Comments

[Roots: *Heterobasidion annosum* not recorded – see p.220]. The authors have found no other useful information on the pathology of *Tetradia*.

THUJA – Thujas

■ **Roots, root collar, butt**

Honey fungus killing 145
Root and butt rots: *Heterobasidion annosum, Hypholoma fasciculare, Armillaria* 226

■ **Bark, cambium of stem, branches, twigs**

Amylostereum canker-rot 64
Coryneum canker 107

■ **Foliage (leaves, shoots, buds), flowers, fruit**

Kabatina shoot blight 152
Aphid damage 109
Scale insects 249
Mined shoots 151
Drought damage 112
Foliage bronzing 130
Urine damage 93

See also Comment below

■ **Structural and water-conducting wood**

Amylostereum canker-rot 64

Comments

The lower foliage of *Thuja plicata* is occasionally speckled brown when scattered scale leaves are killed by the fungus *Didymascella thujina* (keithia disease). Brown then black, round, cushion-like fruit bodies develop on the dead leaves, later falling out to leave conspicuous cavities. This is a serious nursery disease but on outplanted trees and hedges no control is necessary or available.

THUJOPSIS – Thujopsis

Comments

[Roots: *Heterobasidion annosum* not recorded – see p.220]. The authors have found no other useful information on the pathology of *Thujopsis*

TILIA – Limes

■ **Roots, root collar, butt**

Phytophthora killing	206
Root and butt rots: *Ganoderma adspersum, G. applanatum, Pholiota squarrosa, Ustulina deusta*	226

■ **Bark, cambium of stem, branches, twigs**

Tilia × euchlora weeping canker	269
Scale insects	149,251

■ **Foliage (leaves, shoots, buds), flowers, fruit**

Leaf spots	161
Mite damage	172
Leaf galls	133,134
Leaf mines	158
Leafhopper damage	157
Slugworm damage	195
Honeydew and Sooty moulds	144
Drought damage	112
Salt damage	93

■ **Structural and water-conducting wood**

Verticillium wilt	261
Mistletoe	104

Comments

Varying numbers of dead twigs and branchlets are commonly present in the crowns of large *T. × europaea*. Whether there is one universal cause for most such damage is unknown so each case requires separate consideration. Root problems, drought (see p. 115) and bark disease (though none is known) are all possibilities.

[Roots: *Heterobasidion annosum* not recorded – see p.220]

TORREYA – Nutmeg Trees

Comments

[Roots: *Heterobasidion annosum* not recorded – see p.220]. The authors have found no other useful information on the pathology of *Torreya*.

TRACHYCARPUS – Chusan Palm

■ **Foliage (leaves, shoots, buds), flowers, fruit**

Wind tatter	176

Comments

[Roots: *Heterobasidion annosum* not recorded – see p.220]

TSUGA – Hemlocks

■ **Roots, root collar, butt**

Honey fungus killing	145
Root and butt rot: *Heterobasidion annosum, Armillaria*	226

SECTION **2**

■ **Foliage (leaves, shoots, buds), flowers, fruit**

Lime induced chlorosis	174

See also Comments below

Comments

In North America, where it is indigenous, *Tsuga heterophylla* has proved to be one of the most sensitive tree species to sulphur dioxide (SO_2) (see p.96).

ULMUS – Elms

■ **Roots, root collar, butt**

Honey fungus killing	145
Root and butt rots: *Rigidoporus ulmarius, Ganoderma spp,*	
Armillaria	226
Phytophthora root disease	206

■ **Bark, cambium of stem, branches, twigs**

Coral spot fungus	106
Scale insects	149
Corky wings	103

■ **Foliage (leaves, shoots, buds), flowers, fruit**

Leaf spots	161
Leaf galls	133
Leafhopper damage	157

■ **Structural and water-conducting wood**

Dutch elm disease	116
Bacterial wetwood	74
Stem rots: *Polyporus squamosus,*	
Pleurotus species	226
Twig shedding – see Comment (3) under *Populus*	

UMBELLULARIA – Californian Laurel

■ **Roots, root collar, butt**

See Comments below

■ **Bark, cambium of stem, branches, twigs**

Nectria canker	185

■ **Foliage (leaves, shoots, buds), flowers, fruit**

Late (spring) frost	126

Comments

The Californian laurel is said to be more liable to windthrow than any other tree in California.

[Roots: *Heterobasidion annosum* not recorded – see p.220]

ZELKOVA – Zelkova

■ **Structural and water-conducting wood**

Dutch elm disease	116

Comments

[Roots: *Heterobasidion annosum* not recorded – see p.220]

SECTION 3

THE DESCRIPTIONS OF PESTS, DISEASES AND DISORDERS

ARE ARRANGED IN ALPHABETICAL ORDER
OF THE COMMON NAME AS LISTED OVERLEAF

CONTENTS

SECTION **3**

EXPLANATORY NOTES TO THE HEADINGS IN THE DESCRIPTIONS WHICH FOLLOW

THE SUBJECT OF THE DESCRIPTION *(usually the common name of the pest, disease or disorder)*	The causal agent: *(the type of organism and its Latin names with common synonyms; or the non-living agent)*

Damage Type: The principal and direct effect of the damaging agent on the tree (not the secondary symptoms).

Symptoms & Diagnosis: The symptoms most likely to have first drawn attention to the condition and which are evident from no more than a visual examination.

Confirmation: Further symptoms of the condition which provide (sometimes on their own, sometimes only in conjunction with the other symptoms and circumstances of the case) extremely strong evidence of the cause. These are not immediately evident and may require more than a mere visual examination – e.g. the use of a hand lens, or the need to cut into the tree or to expose roots.

Additional indicators: Circumstances in which the condition could be expected to arise, or features of it which are not invariably present. N.B: The diagnosis does not depend on the presence of these additional indicators – it is merely reinforced by them.

Caution: Other pests, diseases or disorders which produce similar symptoms and should, therefore, also be considered.

Status: The frequency with which the condition arises or has been noted and its distribution in Great Britain.

Significance: The importance of the condition in relation to the tree's growth and its amenity function; the implications for the tree's future and for surrounding trees; inferences to be drawn from the presence of the problem.

Host Trees: Tree genera or species liable to the condition. Particularly susceptible species are listed where known if the information is of practical value.

Resistant Trees: Tree genera or species not liable to or not much damaged by or rarely affected by the condition. (If implicit in the list of Host Trees given, this heading is omitted.)

Infection & Development/Reason for Damage/Life Cycle: The life cycle of the pest or disease organism and the development of the condition; or the reason that the abiotic agent gives rise to the disorder.

Control: (To be read in conjunction with Section 4, Control). Whether control is necessary; effective and practicable control measures. Even if effective control measures are known, these are given only if the condition is likely to warrant control in normal circumstances. Because pesticides regulations and the availability of pesticides changes, it will not necessarily be possible to put the recommendations into practice.

Remarks: Miscellaneous information including a brief history of the discovery and origins of the disease or pest.

Further Reading: Fuller, readily available accounts of the condition. For the complete reference refer to the References in Section 6. Other accounts often appear in some of the text books listed in the same list of References and Main Literature Sources.

SECTION **3**

AMYLOSTEREUM CANKER-ROT

Fungus: *Amylostereum laevigatum*

Damage Type: Perennial bark killing and wood rotting.

Symptoms & Diagnosis: One or more dead or dying (fading, yellowing or brown) branches are evident (Fig.17); elongated, sunken stem cankers (Fig.18), perhaps barkless and decayed, may also be present.

Additional indicators: The tree bears pruning wounds or has been topped; the dead branches are aligned more or less one above the other (coincident with a dead bark strip).

Confirmation: The dying branches are found to arise from a narrow strip of dead bark running up and down the stem (Fig.20), usually linked to a pruning stub, topping wound or dead root. The wood in infected stems is stained orange or yellowish-brown (Fig.19) and the wood behind the dead bark may be decayed. Thin patches of whitish mycelium (fruit bodies) may appear on the dead bark here and there, especially near the base of the tree (Fig.21).

Caution: Lightning (p.170) sometimes kills narrow strips of stem bark and the associated branches, but roots are rarely involved.

Status: The fungus is common and widespread on Taxus and Juniperus but whether it causes the disease in them is uncertain. Since its discovery, the disease has been found only a few times on Thuja and Chamaecyparis.

Significance: Rarely fatal because of its slow lateral spread in the stem, but the death of branches can seriously disfigure the tree.

Host Trees: Thuja occidentalis, T. plicata, Chamaecyparis lawsoniana. See also Status, above.

Infection & Development: Observations and experimentation suggest that

Fig. 17

Fig. 18

Fig. 19

Fig. 20

spores infect pruning wounds if these are large enough to expose a substantial amount of sapwood. The fungus grows in the wood, much faster up and down than laterally, and causes a light brown, fibrous, dry decay. If it reaches the cambium, this is killed and a long, narrow, sunken, perennial canker forms, widening only very slowly. The inconspicuous, whitish patches which are the fruit bodies may form on the dead bark here and there. After some years, the dead bark cracks and falls away to reveal the decayed wood. Branches arising from the dead bark are killed but whole trees rarely die.

Control: *Therapy:* None possible as by the time symptoms appear the fungus will be too deeply entrenched for even excision to be feasible.
Prevention: Thuja and Lawson cypress (and yew) are often pruned hard without subsequently dying back. The risk of infection therefore seems to be too slight to warrant routine preventative measures. Were these required, it is likely that the treatment of pruning wounds with a wound paint effective against the Silver leaf fungus (p.254) would succeed against this fungus too. [Take note of current Pesticides Regulations – see p.288]

Remarks: The disease was first described and the cause identified in the late 1970's, in England. This account is based entirely on unpublished work by R. G. Strouts.

SECTION **3**

Fig. 21

ANTHRACNOSE OF LONDON PLANE

Fungus: *Apiognomonia veneta*
(*= Gnomonia veneta; G. platani*)
Anamorph: *Discula platani*
(*= Gloeosporium nervisequum*)

Damage Type: Bud, shoot and bark killing; leaf-vein and petiole killing; defoliation (Fig.22).

Symptoms & Diagnosis: *Syndrome A:* In spring, scattered shoots (perhaps in large numbers) have suddenly wilted and died, as if frosted (Fig.23). *Syndrome B:* In spring, scattered buds or whole twigs have failed to flush, or have flushed weakly and then died. *Syndrome C:* In early summer, leaves, substantially green, are falling, perhaps leaving the tree virtually leafless.
Additional indicators: Many planes in the locality are affected. Scattered old dead twigs and small bark cankers are present.
Confirmation: Syndrome A: is unlikely to be anything other than anthracnose. *Syndrome B:* a patch of dead bark surrounds each dead bud (Fig.25); a band of dead bark girdles the base of the unflushed portion of twigs; this band of bark (often after the whole twig has died) is covered in pimples (Discula fruit bodies) (Fig.24). *Syndrome C:* Brown patches extend along some of the main veins (Fig.26); petioles are dead.
Caution: Gross symptoms might be confused with de-icing salt damage (p.93).

Status: Widespread. Common on certain London plane clones in outbreak years.

Significance: Temporarily alarming and disfiguring but even severely defoliated London Planes recover by mid-summer.

Host Trees: Platanus occidentalis is very susceptible and suffers severe dieback from repeated attacks; P. orientalis is fairly resistant; the susceptibility of P. × hispanica varies according to the clone.

Infection & Development: *Leaf blight:* In warm, wet spring weather, spores from killed twigs infect young leaves. The fungus kills leaf veins (Fig.26) and may spread down the petiole into the twig, causing the leaf to fall.
Bud and twig blight: During the summer the fungus remains dormant in the twigs but during warm periods in winter will resume activity and kill patches of bark. Buds (Fig.25) or twigs (Fig.24) girdled in this way do not flush, or flush and quickly die.
Shoot blight: sometimes little damage is done in winter but then, if, in the two weeks following flushing, temperatures are sufficiently low to inhibit shoot growth but high enough for relatively good fungal growth, the fungus may spread into the bases of the new shoots and kill them (Fig.23).

Control: *Therapy:* Unnecessary.
Prevention: If resistant clones of P. × hispanica become available (some are currently marketed in the USA), their use or the use of P. orientalis, where this is a suitable choice in other respects, would overcome the problem.

Remarks: The disease was first described in 1815, in England, on P. occidentalis, and is in all probability the reason for the scarcity of that tree there now. The fungus (anamorph) was described in 1848. The link between fungus and disease was demonstrated in 1905, in Germany, and the teleomorph described.

Further Reading: Strouts, 1991a. Gibbs, 1983.

Fig. 22

Fig. 25

Fig. 23

Fig. 26

Fig. 24

SECTION 3

ASH BUD MOTH   Insect: *Prays fraxinella*
(= Prays curtisella)

Damage Type: Hollowing out of buds; removal of bark at base of shoots, sometimes girdling and killing them.

Symptoms & Diagnosis: In May the new shoots wilt (Fig.27), die and become black. The terminal buds fail to flush and there may be some dieback of small twigs.
Confirmation: Look for entrance/exit holes in unflushed buds and for hollow buds. Silken webs and frass are often present.
Caution: Holes and short tunnels in twigs are also caused by overwintering ash bark beetles (Hylesinus spp. – not further described here).

Status: Widespread and probably common although not often reported.

Significance: The form of the tree is affected. The loss of terminal buds and shoots causes forking or asymmetrical growth when a lateral bud subsequently develops. P. fraxinella affects trees of all ages but the damage is most serious on young ash. There maybe a connection between the presence of P. fraxinella and ash dieback (see p.37).

Host Tree: Fraxinus excelsior.

Life Cycle: The adult moth flies in June and July and lays tiny eggs singly on the underside of the leaves. These hatch in 5-8 weeks and the larvae mine straight into the leaves where they remain for about two weeks during late September or early October. Before leaf fall the larvae move onto the stem and either mine a bud or tunnel into the bark. They feed and overwinter in these places but may change the feeding site several times. In spring the larvae feed on the flowers before mining in the pith at the centre of new shoots. They are

Fig. 27

fully fed by the end of May when they pupate in a loose web in a twig crotch.

Control: None tried. On small trees the use of an insecticide such as diflubenzuron in early October may be effective. [Take note of current Pesticides Regulations – see p.288]

Remarks: This may be more common on trees stressed by root damage: farmland ash close to deep ditches tend to have more buds damaged than trees not alongside ditches.
Small shoots are sometimes killed during the winter when they are girdled by bark-tunnelling larvae.
Buds which survive minor insect damage are sometimes then killed by the fungus Nectria (p.185) which has gained entry through the feeding wounds.

BACTERIAL CANKER (or KNOT) of ASH

Bacterium: *Pseudomonas syringae* ssp. *savastanoi* pv. *fraxini*

Damage Type: Bark proliferation and killing.

Symptoms & Diagnosis: Erumpent or sunken patches of very roughened, dark, thickened, cracked bark are seen on branches or stems, perhaps numerous and up to the diameter of the branch in width (Fig.28). Some branches are dead (Fig.29).
Additional indicators: The underlying wood is usually not exposed, but if it is, no clear target canker is apparent; the excrescences are usually not centred on a bud or twig.
Caution: Compare Nectria canker (p.185).

Status: Widespread, but usually only a small proportion of any collection of ash trees is affected.

Significance: The low incidence of the disease prevents it from assuming any importance in the cultivation of ash.

Infected trees tend to bear many separate infections so become very disfigured and spoilt as timber, but they continue to grow, with the occasional death of a branch, despite the disease.

Host Trees: Fraxinus excelsior and its varieties.

Resistant Trees: The disease has not been reported on other Fraxinus species. Work in 1980 confirmed that clones of F. excelsior vary greatly in susceptibility.

Infection & Development: This has still not been fully elucidated. The bacteria persist for many years in the thickened, infected bark, slowly spreading into surrounding healthy bark, apparently during each growing season. Infection results in an increase in the number and size of cork cells. The swelling, after some years, causes the normal smooth bark to split and the

SECTION 3

Fig. 28

Fig. 29

infected area assumes the characteristic roughened surface. Often, little cambium is killed, the damage consisting then just of the gall-like excrescences, when the disease is not strictly a 'canker' (Bacterial knot is a suggested alternative name). Sometimes, however, cambial killing is extensive, and subsequent alternating growth and killing of surrounding healthy tissue results in the formation of a flattened or sunken irregular, open canker, though still with a surround of the abnormally thick, rough, corky bark.

In spring or early summer, the bacteria ooze out of the infected bark from young cankers as a yellowish slime. Neither the usual points of infection nor the way in which the bacteria reach these are known, but mechanical bark injuries are certainly vulnerable. New infections are usually on young branches and on trees over about ten years old.

Control: None required

Remarks: It seems likely that the incidence of the disease is low because few susceptible individuals arise among the seedlings which make up our ash population.

The disease was described in England in 1855. The bacterial cause was elucidated in 1889, in France and experimentally confirmed in 1932, in the USA.

BACTERIAL CANKER OF CHERRY

Bacterium: *Pseudomonas syringae* pv. *morsprunorum*

Damage Type: Bark killing.

Symptoms & Diagnosis: In spring, scattered dwarf shoots, twigs, branches or whole young trees have failed to flush, or have flushed then wilted and died, or have flushed with yellow leaves (which die later in the season) (Fig.30).
Confirmation: Cutting reveals large areas of dead and dying bark on affected branches, probably girdling them (Fig.31). Amber coloured gum exudes copiously here and there from bark of affected branches or stem (sometimes from branches showing no foliar symptoms) (Fig.32).
Caution: Cytospora laurocerasi (p.110) causes similar symptoms on Prunus laurocerasus. Monilinia laxa (p.85) kills twigs and causes gumming in Prunus but rarely on wood older than two years.

Status: Common and widespread.

Significance: A very disfiguring, sometimes fatal disease.

Host Trees: Prunus avium and its ornamental and fruiting varieties; vari- ous other Flowering cherries including P. cerasifera and its varieties but information on relative susceptibility of the

Fig. 30

Fig. 31

Fig. 33

Japanese cherries is lacking. Also on P. cerasus (sour cherries), P. domestica (plums), P. amygdalus (almond), P. armeniaca (apricot), P. persica (peach).

Fig. 32

Resistant Trees: It is not clear from the literature whether this disease occurs on evergreen Prunus species.

Infection & Development: During summer, the bacteria live on the leaves. Brown spots develop, crack around their margins and fall out to leave small, round holes – the so-called 'shot-hole' symptom (Fig.33). In autumn, in all species but plum, the bacterium infects the bark via the scars left by the falling leaves. In all species, infection can also occur through bark injuries (natural cracks, as in crotches, frost cracks, pruning wounds etc.). During winter, the bacteria remain inactive but in early spring they spread rapidly through the bark. Some branches or the stem may be girdled and killed while non-girdling patches of dead bark show up in later years as sunken cankers. During summer the bacterium in the bark dies out but by this time the new season's leaves are infected and carry the bacterium through the summer.

Control: *Therapy:* None available: by the time symptoms are noticed, the damage has been done. The excision of dead bark is unnecessary as infections do not extend after the first year.

Prevention: Prune only in June, July or August when the heavy gum production minimises the risk of wound infection. Wounds made outside this period should be treated with a suitable protectant, e.g. one based on octhilinone. Otherwise, prevention entails the annual application in autumn of a bactericide to reduce the population of bacteria on leaves – not usually practicable or justifiable in ornamentals but routine in orchards, where resistant varieties and rootstocks are also used. [Take note of current Pesticides Regulations – see p.288]

Remarks: The strains of the bacterium which infect plum do so through bark wounds, not leaf scars. Large cankers are more common on plum than on cherry.

The shothole symptom can also be caused by Blumeriella jaapii (see p. 166) and other less common fungal and viral diseases. The presence of bacterial shothole does not indicate that the tree is suffering from the canker phase of the disease nor that it will necessarily ever contract it.

The bacterial nature of the disease was confirmed in the 1920s, in England.

BACTERIAL CANKER OF POPLAR

Bacterium: *Xanthomonas populi*
(= *Aplanobacter populi*)

Damage Type: Bark killing.

Symptoms & Diagnosis: Dead or dying twigs and branches in large or small numbers are evident, scattered through the crown (Fig.34); or small or large bark cankers, either roughened and irregular or smooth and even, are

Fig. 34

Fig. 35

Fig. 36

Fig. 37

present on branches or stem (Figs.35, 36).

Confirmation: Cutting into bark of dying or recently killed branches and twigs reveals girdling patches of dead bark, often centred on dead buds or twigs. Recently killed bark, especially on young twigs and in damp weather in spring, exudes a whitish mucilaginous bacterial slime. (For diagnosis in dry weather, place girdled or cankered twigs in water for an hour or two to induce slime production (Fig.37).)

Status: Widespread and very common.

Significance: A dominant problem in the cultivation of poplars for amenity and timber in Great Britain and parts of NW Europe.

Host Trees: Has been recorded on Salix but virtually confined to Populus. Of the poplars more commonly used or more suitable for amenity plantings, the following are very susceptible: P. alba 'Pyramidalis', P. candicans (Balm of Gilead), P. candicans 'Aurora', P. × generosa, P. 'Heidemij', P. 'Oxford', P.

'Rap', P. 'Regenerata', P. 'Rochester', P. tremula (aspen).

Resistant Trees: Many of the most resistant poplars have been bred for timber production and so are not necessarily suitable for a particular amenity purpose. The following are very resistant: P. 'Balsam Spire', P. nigra, P. nigra var. betulifolia, P. nigra 'Gigantea', P. nigra 'Italica', P. nigra 'Plantierensis', P. nigra 'Vereecken', P. trichocarpa 'Fritzi Pauley', P. trichocarpa 'Scott Pauley'. The hybrid 'UNAL' clones, introduced from Belgium in 1985, are also very resistant (see Jobling, 1990).
The following rarely suffer more than the death of a few small twigs and branches: P. alba, P. alba 'Raket', P. 'Eugenei', P. 'Gelrica', P. 'I-78', P. 'Robusta', P. 'Serotina', P. 'Serotina Aurea'.

Infection & Development: The bacteria overwinter in bark cankers. At about bud burst, bacteria ooze out of bark cracks in a white mucilage and are distributed in wind and rain; infection is mainly through stipule scars or scars left where leaves have been removed prematurely by wind or fungal disease, or through hail- or insect-damaged bark. Bark killing occurs in dormancy; some occlusion occurs each growing season but the lesion gradually enlarges to form a 'target' canker or a very rough, irregular one, depending on the tree species or variety. One-year-old twigs are often killed after one year; larger members may be girdled only after many years, depending on host resistance. Axially elongated cankers form when the bacteria spread along cambial larval galleries of the agromyzid fly, Phytobia cambii.

Control: *Therapy:* If the tree is remote from other diseased trees, excision of all infected parts when the symptoms first appear should postpone severe disease. *Prevention:* Ideally, use resistant trees.

If using susceptible trees, plant well away from or first remove diseased trees.

Remarks: Even susceptible trees will often escape the disease for many years if grown far removed from diseased trees, and although P. candicans 'Aurora' is very susceptible, it is likely to be little damaged if it is hard pruned annually (as is often done to induce an abundance of its attractive foliage). The disease was first reported from France, in 1884. It appears not to have been known in this country in 1923 but was common in England by 1934. The disease had been attributed to the wrong bacterium for some time until, in 1958, work in France showed a previously undescribed species to be the cause.

Further Reading: Jobling, 1990.

BACTERIAL WETWOOD

Bacteria: *A mixture of various species*

Damage Type: Watery emissions, branch killing.

Symptoms & Diagnosis: An initially colourless to brown, smelly, watery liquid seeps, runs or bubbles out of bark cracks or wounds on trunks or limbs (Fig.38). This soon darkens (often greyish, sometimes black or reddish), and accumulates in an unsightly streak or patch below and around the wound (Fig.39). The deposit may be lumpy and several millimetres thick. When it is moist it is slimy, when dry it may appear chalky.
Additional indicators: The flux issues from a long, vertical crack penetrating into the wood and bounded by two thick callus rolls (forming a seam-like scar). Some branches are dead or carry yellowed, scorched or wilted foliage.
Confirmation: If a hole is drilled into

the infected wood, copious amounts of liquid may flow or even spurt out, perhaps accompanied by gas. The flux is slightly alkaline (about pH 7.5) whereas healthy xylem sap is slightly acidic (about pH 6.5).
Caution: Watery fluxes, some malodorous, some frothy and with a pleasantly fermenting smell, are also caused by various other bacteria, often around small (e.g. insect) wounds or small patches of dying bark. Usually these are of little consequence, but see Beech bark disease (p.77), unexplained Weeping canker of Caucasian lime (p.269), and Fire damage (p.122). Seepage of a gummy fluid usually indicates that the bark is dead (see e.g. Phytophthora bleeding canker (p.202) and Bulgaria canker (p.91)). Exudations from the lower stem may indicate root disease.

Fig. 38

Fig. 39

SECTION 3

Status: The fluxing is common and widespread; associated death of branches is much less common.

Significance: Potentially harmful, occasionally fatal, but often the fluxing is not accompanied by dieback.

Host Trees: Many broadleaves. Commonest on Aesculus, Liriodendron, Ulmus and, according to the literature, Populus. Also on conifers.

Infection & Development: It is thought that the bacteria enter the xylem via the roots from the soil. They are able to multiply vigorously despite the lack of oxygen in the xylem but in so doing produce much methane. Water contaminated with toxins produced by the bacteria accumulates in the infected wood and, by the pressure of the gas, is either forced into the transpiration stream, causing the wilt symptoms, or harmlessly to the outside through cracks in the wood, to be discoloured by the activities of yeasts.

Control: As is clear from the above, the cracking and fluxing should be regarded as the operation of a safety valve so no attempt should be made to stop the fluxing. If considered objectionable, the deposit could be scrubbed off now and then. Continuing dieback cannot be arrested.

Remarks: The anaerobic conditions in bacterial wetwood prevent its decay by fungi. For this reason and because bacterial wetwood is so common yet rarely results in overt disease, it has been suggested that it has evolved as a protection against fungal decay.

Further Reading: Rishbeth, 1982

BALSAM WOOLLY APHID   Insect: *Adelges piceae*

Fig. 40

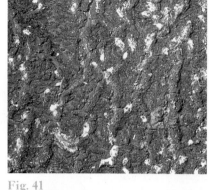

Fig. 41

Damage Type: Swellings on shoots and crown dieback.

Symptoms & Diagnosis: Two distinct symptoms are caused by this insect, though both may occur on the same tree: (i) conspicuous swellings are present around buds and (particularly on Abies procera) around the male flowers (Fig.40); (ii) the crowns of trees of any size may be sparsely foliated, lacking in vigour and dying back.
Confirmation: The main stems and branches of Abies may be covered in whitish, fluffy, waxy wool (Fig.41), there may be resin bleeding (Fig.42) and a close examination will reveal the presence of dark greyish aphids (adelgids). The stem surface may show depressed areas.

Status: Widespread in England, Wales and Scotland.

Significance: Colonies of this adelgid occur mainly on the stem causing severe dieback; some trees may be killed. The feeding activity of the insect affects cell division in the cambium and leads to the development of 'rotholz', dense and brittle growth rings which disrupt the

Fig. 42

transpiration flow and result in progressive crown dieback.

Host Trees: Abies alba is the original European host, but A. piceae also occurs frequently on stems of A. grandis and A. procera; also on A. balsamea,

A. amabilis, A. lasiocarpa, A. cilicica. Both gouty growths on the shoots and wool on the stems occur on A. balsamea, A. amabilis, A. fraseri, A. sibirica, A. lasiocarpa var arizonica and A. koreana.

Life Cycle: A. piceae overwinters as a nymph. Eggs are laid in April or May and a succession of wingless generations continues throughout the summer on stems, shoots and in shoot axils. Winged forms are extremely rare.

Control: Fell severely infested trees that are dying back during the winter (November-March); at all other times there is a risk of infesting further trees as the A. piceae crawlers will be disturbed when the tree falls and will then be dispersed by the wind, probably for considerable distances.

BEECH BARK DISEASE	Insect: *Cryptococcus fagisuga* Fungus: *Nectria coccinea*

Damage Type: Bark killing; wood decay by secondary invaders.

Symptoms & Diagnosis: *Early stage:* Small blackish spots appear, scattered over the stem bark where a sticky liquid has seeped out and dried (Fig.43). *Later stage:* Patches of stem bark, perhaps extensive, have cracked or fallen off: the crown of the tree is thin and the foliage yellowish.

Latest stage: A tree with much of its stem bark missing or falling off has died; or the stem of a tree with a thin, yellow crown has decayed and snapped. *Additional indicators:* A large area of bark has been covered in a white, fluffy material (Cryptococcus colonies) for some years prior to its death (Fig.129). This may still be present, though perhaps only as scattered woolly spots or patches.

SECTION 3

Fig. 43

Fig. 44

Fig. 45

Confirmation: Early stages: probing the black 'tarry spots' reveals that each marks a small patch of dead or dying bark (Fig.44), or that the whole area of spotted bark is dead. *Later stages:* Dead bark may be reddened in large or small patches with innumerable, pinhead-sized pimples (Nectria fruit bodies) (Fig.45). The major roots are the last part of the tree to die.

Caution: Severe drought stress can permit the invasion of bark by Nectria coccinea just as infestation by the coccus can. The two diseases then develop similarly so can be hard to differentiate, though in practice the precise diagnosis is by then academic.

Status: Occasionally kills trees over about 20 years of age but much less common in trees over about 80 years old. Widespread.

Significance: Although potentially fatal, in specimen trees it can be guarded against and in woodlands it can be tolerated.

Host Tree: Fagus sylvatica.

Infection & Development: Prolonged and heavy infestation of bark by the Felted beech scale (coccus) (Cryptococcus fagisuga – see p.120) renders it open to invasion and killing by the weakly pathogenic fungus Nectria coccinea. The initially scattered, small fungal colonies spread radially and coalesce to produce extensive dead areas. Lacking the protective covering of live bark, the wood is open to attack by ambrosia beetles (especially Trypodendron domesticum) and, via the beetles' tunnels or direct, to wood rotting fungi – commonly Bjerkandera adusta. As the bark dies, so does the coccus.

Control: *Therapy:* Once symptoms are evident, it is too late for effective control measures. In commercial woodland, infected trees of marketable size should be utilized before decay can set in . Specimen trees should be monitored for decay development (see Section 5, Decay and Safety). Some recover spontaneously, the small dead bark patches callus over and little decay develops.

Prevention: Destroy or prevent the development of heavy coccus infestations by scrubbing colonies with a mild detergent or by applying a tar oil winter wash as recommended for fruit trees. [Take note of current Pesticides Regulations – see p.288]

Further Reading: Lonsdale and Wainhouse, 1987.

BEECH LEAF-MINING WEEVIL

Insect: *Rhynchaenus fagi*

Damage Type: Removal of internal leaf tissues.

Symptoms & Diagnosis: Brown blotches occur in the apical half of the leaf, reaching the margin (Fig.46); a narrow brown streak leads from the midrib to each blotch. There may be several blotches per leaf, which may be distorted. Damage occurs in May/June and remains visible until leaf fall (Fig.47). *Confirmation:* The blotches are mines which are translucent if held up to the light. Small holes (so-called 'shot-holes') may also be present on blotched or nearby unblotched leaves.

Status: Widespread and in some years very abundant, especially in woodlands, wherever the host grows.

Significance: Severe attacks spoil the appearance of ornamental trees and sometimes of hedges. Leaves damaged in spring remain on the tree throughout the summer giving affected Fagus a shabby or unthrifty appearance.

Host Trees: Mines occur only on Fagus sylvatica and its cultivars. Shot holes also occur on other trees and bushes in the vicinity of affected Fagus, especially on Crataegus in April and Rubus idaeus (raspberry) in summer.

Life Cycle: Adult weevils overwinter in bark crevices, litter, etc; in spring they become active early and feed on the beech leaves as soon as these flush, creating the 'shot holes'. Eggs are laid near the midrib in early May. Each larva mines a narrow corridor which then widens into an irregular blotch towards the leaf margin. The larvae pupate in the mines from which the new generation of adults appear in June. These move away and feed during the summer by chewing small 'shot' holes in the leaves of a variety of trees

Fig. 46

Fig. 47

and bushes (see Host trees above). There is only one generation per year.

Control: None necessary.

SECTION 3

BEECH WOOLLY APHID
Insect: *Phyllaphis fagi*

Damage Type: Foliage discoloration and distortion; shoot dieback on hedges.

Symptoms & Diagnosis: The leaves are curled downwards and become puckered and distorted with brown edges. On hedges and on transplants in treeshelters, some shoot dieback may be present.

Confirmation: Aphids and waxy wool mixed with honeydew are on the underside of the leaves (Fig.48). Honeydew also drops onto the top surface of the lower leaves and these become blackened with Sooty moulds (see p.144).

Fig. 48

Status: Common and widespread.

Significance: The Beech woolly aphid causes a significant loss of growth in young plants and may be a contributory cause of some deaths in tree shelters. On hedges the aphids and honeydew are sometimes a considerable nuisance, especially on regrowth following cutting. Sooty moulds and browning are disfiguring. Damage by this aphid is of little significance on large trees.

Fig. 49

Host Trees: Fagus sylvatica and its cultivars; also known on Parrotia persica.

Life Cycle: Aphids are present and breeding from when the tree flushes in spring until just before leaf fall. They feed by sucking phloem sap from leaves. Reproduction slows at the end of June but may resume when the lammas growth appears. Numbers are low in July and August when the aphid apparently enters a summer quiescent period as the leaves mature. Winged aphids appear twice in southern England with peaks in June/July and in October. Further north, flight continues over one long period with an autumn peak. P. fagi eggs overwinter on the bark of twigs near to the buds (Fig.49).

Control: Difficult because aphids are protected by waxy wool and by their feeding position on the underside of curled leaves. On hedges use a systemic insecticide soon after flushing. Alternatively, on nursery stock, the use of a fatty acid spray to break down wax, followed immediately by a pyrethroid insecticide such as cypermethrin, has given good results. Also on nursery stock, the survival of large numbers of overwintering eggs can be prevented by using a tar oil winter wash during the dormant season. These treatments prevent eggs being transported on planting stock and hence stop the build-up of

P. fagi populations in tree shelters. There are no current approvals for either of these uses on hedges. [Take note of current Pesticides Regulations – see p.288] No control is necessary or recommended on larger trees.

BLACK FLECK DISEASE OF RED HAWTHORN

Cause: *Unknown*

Damage Type: Not known but appears to originate in the rootstock.

Symptoms & Diagnosis: Not fully elucidated; the following description must be regarded as provisional: a tree, apparently healthy at the end of one summer, may not flush in the following spring; or soon after flushing, leaf margins may go brown, leaves shrivel but remain on the tree, and flower buds fail to open fully; as summer wears on, the abortive flowers die and more and more leaves die and turn brown, though the bark of shoots, twigs, branches and main stem may remain alive until late in the summer, when all foliage and the bark over most of the tree dies.

Fig. 50

Confirmation: Cuts through the bark of the rootstock reveal numerous black (sometimes reddish brown or wine-red) flecks or patches in the outer centimetre of the xylem. The flecks occur in the roots and the stem base but terminate abruptly above in a distinct more or less straight line at the junction between rootstock and scion (Fig.50). Flecks extend further along the grain than across it and vary in size from about a millimetre long and less than a millimetre wide to streaks of up to 3 cm long and 2 mm wide, occasionally larger. They tend to be more numerous towards the root tip, sometimes running together to blacken the whole of the outer wood but never staining the bark. In cross section the flecks are wedge-shaped, narrowing towards the pith and passing through several annual rings or affecting only the outermost. When crown symptoms first appear, the bark of the rootstock is likely to be alive but by late summer will probably have died.

Additional indicators: In some cases, a few dead roots among mostly live ones and patches of dead bark on the stem base as described for Phytophthora root disease have been found (p.206). In one instance Phytophthora was isolated from this dead bark but it is not yet clear whether this is a usual part of the syndrome or was merely coincidental.

Caution: Similar crown symptoms may be induced by Phytophthora root disease (p.206), Honey fungus (p.145) or Fomes root disease (p.220).

Status: To date, known only from fewer than a dozen scattered localities in the southern half of Great Britain and on only one or a few trees in each instance.

SECTION 3

Significance: Unclear, but a potentially fatal disease.

Host Trees: All known cases have been on grafted hawthorns, none known to be white-flowered. Almost certainly, these have all been varieties of Crataegus laevigata (= C. oxyacantha), some certainly 'Paul's Scarlet'.

Infection & Development: Mostly unknown but root malfunction seems to be the cause of the crown symptoms.

Control: None known.

Remarks: The disease has been seen in trees from 10 to 50 years old. The first recorded cases were from Derby in 1985. Next to nothing is known about the aetiology of this disease. Useful information might be gained readily if producers of red flowered hawthorns were to check rootstocks for the distinctive symptoms before and after grafting.

BLACK LINE DISEASE OF WALNUT

Cause: *Cherry leaf roll virus*

Damage Type: Killing of phloem and cambium.

Symptoms & Diagnosis: The tree is growing feebly, the crown is thin, dieback may be apparent and healthy shoots may be sprouting from the rootstock in profusion.
Additional indicators: These symptoms may have been developing over a period of up to about 5 years.
Confirmation: If the outer bark is cut away at the graft line, a black, necrotic band of cambial and phloem tissue several millimetres wide and coincident with the graft union is revealed, extending part or all of the way round the tree (Fig.51).

Status: A rare disease now that Juglans nigra is no longer recommended as a rootstock for J. regia.

Significance: A fatal disease and a good example of successful control by a change in nursery practice. It also suggests a possible explanation for other, otherwise unexplained, cases of 'graft incompatibility' which manifest themselves only after some years of healthy growth.

Susceptible Trees: Juglans regia grafted onto J. nigra.

Infection & Development: Once the J. regia scion is old enough to produce flowers (10 years or so after grafting), it is open to infection by the Cherry leaf roll virus (CLRV), brought in on infected, wind-blown pollen. J. regia is tolerant of the virus and if infected shows no more than some yellowing of leaves as the virus slowly spreads through it. The J. nigra rootstock is, however, hypersensitive to the virus, so that when the pathogen reaches the rootstock, the phloem and cambial cells are killed. This prevents the further downward spread of the virus into the rootstock but the stem is girdled and the J. regia

Fig. 51

scion dies. The rootstock responds by sprouting vigorously.

Control: *Therapy:* None available. *Prevention:* J. nigra should not be used as a rootstock for J. regia.

Remarks: The disease was first noted in the USA in the 1920s; the virus was first described in 1961, in Britain; the connection between the virus and the disease was proved in 1984 in the USA. Many cases of virally induced graft failure are known, and some (e.g. Apple union necrosis) resemble Black line disease in symptomatology. Any deaths of the scions of grafted trees after years of healthy growth are likely to be the result of such diseases.

BLISTER RUST OF WHITE PINE   Fungus: *Cronartium ribicola*

Fig. 52

Fig. 53

Damage Type: Bark killing.

Symptoms & Diagnosis: One or more branches or the top of the tree is dead or dying (Figs.52,53). The portion of the branch or stem between the dying and live parts is exuding resin (and is perhaps also swollen and the bark cracked).
Additional indicators: This disease should be the first suspect where scattered branches die on a five-needled pine.
Confirmation: In early summer, prominent, yellowish-white, sac-like outgrowths, a few millimetres high, appear on the surface of the swollen, cracked areas of bark. These rupture to reveal the orange, powdery contents (aeciospores) (Fig.54).

Status: The disease may occur anywhere in Great Britain though, as it has limited the successful cultivation of susceptible species, it is not common.

Significance: An often fatal disease, and one which is so likely to occur sooner or later on very susceptible trees that, if they are used at all, they should form only a small part of any planting scheme.

Host Trees: Confined to pines which bear their needles in bundles of five (but see Remarks' below); any species may be infected. Those from the New

SECTION 3

World are nearly all highly susceptible (e.g. P. flexilis, P. lambertiana, P. monticola, P. strobus); when attacked, those from the Old World suffer far less damage (e.g. P. cembra, P. peuce, P. wallichiana).

Infection & Development: Airborne basidiospores are produced on Ribes leaves (mainly Black currant) during summer and autumn. These infect pine needles of any age. The fungus grows from the needle into the shoot and becomes established permanently in the bark, spreading each year further down the branch, eventually girdling it. If it reaches the main stem this too may be girdled in time. In early summer, the infected, swollen bark, produces orange aeciospores in small, white, sac-like outgrowths (Fig.54). These spores cannot infect pines but infect leaves of Ribes. Damage (from defoliation) on the Ribes is usually slight. During summer, urediniospores produced on the Ribes reinfect Ribes, while basidiospores are also produced which can infect only pines.

Control: *Therapy:* Growth of the fungus from an infected branch into the stem can be prevented if the branch can be severed below the lower limit of the infection.
Prevention: None available (The alternate hosts (Ribes species) are too widespread and common and spores can be carried too far for their destruction to be helpful).

Remarks: The fungus was first described in 1854, from Estonia,

Fig. 54

though it is thought to have originated in Asia, probably on Pinus cembra, on which it causes little damage. Soon after the North American P. strobus began to be planted widely in Europe in the latter part of the 19th century, the disease became epidemic on it and consequently it was abandoned as a forest species in Europe.

Pinus sylvestris is subject to a similar disease (common in East Anglia and NE Scotland, rare elsewhere) but the causal fungus (Peridermium pini) requires no other plant on which to complete its life cycle. It is virtually confined to this species of pine.

| BLOSSOM WILT, SPUR BLIGHT AND WITHER TIP | Fungus: *Monilinia laxa (and f.sp. mali)* (= *Sclerotinia laxa*) Anamorph: *Monilia laxa* |

Damage Type: Flower, leaf, fruit and young bark killing.

Symptoms & Diagnosis: In full flush, flower trusses hang brown, withered and dead on the tree (Fig.56). All leaves on some spurs (dwarf shoots) are dead but firmly attached (Fig.57). Leaves on the ends of some growing shoots are wilted and shoot tips withered. Soon after flushing, some one-year-old extension shoots are dead (see note regarding P. subhirtella 'Autumnalis' under Host Trees below). Later in the season, fruits hang dead and withered. The amount of damage can vary from a few scattered affected parts to a very large proportion of the whole crown (Fig.58).

Confirmation: Withered flowers smell distinctly sweet. Dead patches of bark extend from some spurs into older wood. In damp weather, small patches of a sweet-smelling, greyish mould are visible on withered fruits. On Prunus species, patches of gum are present at the base of some affected long shoots or on spurs (Fig.55).

Caution: The symptoms, but not their distribution, resemble spring frost damage (p.126).

Pseudomonas morsprunorum (p70) also causes twig death and gumming but does not cause Wither tip, and any blossom death results from the death of the larger branches which bear them.

Status: Common and widespread.

Significance: Can ruin flowering in bad seasons. In ornamentals, much more common on Flowering cherries than on Flowering crabs. An important orchard disease.

Host Trees: Among ornamentals, predominantly a disease of Flowering cherries, less often crabs, but may also occur on pear, peach, nectarine, apricot, plum

SECTION 3

Fig. 55

Fig. 56

and almond as well as cherry and apple trees grown for fruit production. Observations indicate that Prunus sargentii, Prunus × hillieri 'Spire' and P. 'Cheal's Weeping' can be badly damaged. This disease is probably the usual cause of the death of 1-year-old twigs on the winter flowering cherry, P. subhirtella 'Autumnalis'. On this tree the damage does not seem to be associated with the death of flowers. Reports of serious damage on Flowering crabs are scarce.

Infection & Development: The fungus overwinters in the bark of twigs and (except in Malus) in fruits infected in the previous year. In spring, spores produced on these parts, carried in wind, rain or on insects, infect the current year's new flowers and shoots. The fungus grows through and kills these and sometimes spreads into one- or two-year-old bark to girdle twigs.

Control: *Therapy:* None available: by the time symptoms appear, the damage is done.

Prevention: Although usually impracticable, the removal of all affected parts will greatly reduce infection in the following year. Similarly, various fungicides applied at flowering time are very effective, e.g. benomyl, carbendazim, vinclozolin, iprodione. [Take note of current Pesticides Regulations – see p.288]

Remarks: The dramatic and sudden death of flowers often leads to a fear that Fireblight is to blame. Fireblight, however, does not affect Prunus (see p.124). Known in Europe since the late 19th century.

On Malus, the disease is caused by Monilinia laxa f.sp. mali. This form of the fungus does not affect Prunus, nor does the form on Prunus affect Malus.

Fig. 57

Fig. 58

BOX SUCKER   Insect: *Psylla buxi*

Damage Type: Bunching and distortion of terminal leaves.

Symptoms & Diagnosis: In spring the young leaves curl as the buds open and cabbage-like galls up to 2cm diameter result at the tips of new shoots (.59 arrow). There are white or bluish curly waxy woollen threads present which, in heavy infestations, are so abundant that they fall and accumulate in masses on the ground below affected trees or hedges. There is also honeydew and sooty mould present (p.144). By June the leaves below the galls on the new shoots may become distorted or blistered.

Fig. 59

Status: Locally common throughout Great Britain.

Significance: The galls are particularly disfiguring on hedges as they remain long after the damage occurred.

Host Trees: Buxus sempervirens, B. balearica.

Life Cycle: Eggs are laid in August and overwinter on the shoots. In spring the young nymphs feed on the growing tips of shoots within the protection of the developing galls. Adults of P. buxi appear from the end of April until August.

Control: None recommended. The clipping that box hedges normally receive removes most of the visible galls.

SECTION **3**

BUD BLIGHT OF SPRUCE	Fungus: *Gemmamyces piceae* (= *Cucurbitaria piceae*) Anamorph: *Camarosporium strobilinum* (= *Megaloseptoria mirabilis*)

Damage Type: Bud killing.

Symptoms & Diagnosis: Un-flushed buds, dead twigs, or irregularly (instead of symmetrically) branched and crooked twigs appear through the crown. Severely affected trees look 'moth-eaten' (Fig.60).

Confirmation: Buds which should have flushed but have not are dead; some of these began to extend before dying and are enlarged and perhaps

Fig. 60

Fig. 61

strongly curved (Fig.61). Many of the dead buds are covered in a black, finely warty encrustation (the fruit bodies of the causal fungus) (Fig.62 – much magnified).

Status: Almost all British records have been from forests in the north and west (principally in river valleys) but a severely damaged Picea asperata is known as far east as Hampshire. Given the nature of the symptoms, the disease must often be overlooked. Infections can be locally widespread but may involve only a few buds on each tree.

Significance: The disease can severely disfigure the more susceptible species but only very localized cases of such severity are known.

Host Trees: Most records in this country are from the common Norway spruce, Picea abies, but damage on these has been slight. Severe damage occurs on P. engelmannii and one severely damaged P. asperata is known. Other susceptible species are P. pungens (including the commonly planted variety glauca) and P. glauca. A 1932 report lists the disease on Picea sitchensis in Scotland but no other cases have come to light since then and, in view of the prominence of this species in forestry in areas where the disease occurs on P. abies, the record seems dubious. On the Continent it also occurs on Abies alba.

Infection & Development: The fungus is confined to the buds. In June, the sticky spores, dispersed in blown and dripping rainwater (and perhaps on the beaks and feet of birds) infect still-developing buds which are not yet protected by resin. By the following spring, a mat of mycelium has formed which holds the cap of bud scales in place, preventing flushing or distorting the expanding shoot. Killed unflushed or partly flushed buds develop a black warty crust of fruit bodies from which, in wet weather, ooze white masses of conidia. In the second spring after infection, yellow masses of ascospores ooze from the same black encrustations.

Control: *Therapy:* None available.
Prevention: Should a badly diseased tree require replacing, it would, on general grounds, be sensible to replant with a less susceptible species.

Remarks: First described in 1909, from Scotland, on P. pungens var. glauca. The information on the disease's infection and development given here has been supplied by Dr. A. T. K. Corke from his unpublished thesis (1955).

Fig. 62

BUD PROLIFERATION OF RED HORSE CHESTNUT

Cause: *Unknown*

Damage Type: Proliferation of buds and dwarfed shoots on swellings on stems and branches.

Symptoms & Diagnosis: In winter, conspicuous clusters of buds are evident on branches or stems (Fig.63), or in summer crowded clusters of weak shoots bearing dwarfed leaves are noticeable. The clusters may consist of just a few buds or shoots on a smooth, blister-like bark swelling a few centimetres across or of a very large number packed close together over a swelling half a metre or more across and 15 cms thick. The whole structure may be alive or some buds, twigs and bark may be dead, or the whole may be dead and it and the underlying wood decaying. Some swellings are much greater in diameter than the branches bearing them and may lack buds and shoots.

Trees may bear any number of the structures. Often they occur around the graft line or round pruning or other wounds. Many, however, develop in the absence of evident wounds.

Confirmation: This disorder is so distinctive and singular that it may be used to distinguish A. × carnea from other Aesculus species with a high degree of reliability.

Status: Common and widespread in England. Probably equally common in the rest of Great Britain wherever Aesculus × carnea is grown.

Significance: For many years merely disfiguring, but if the galls die and the underlying wood is invaded by wood-rotting fungi the tree may become unsafe.

Host Trees: There appear to be no records on any tree but the hybrid red-flowered Horse chestnut, Aesculus × carnea, and there is no mention of the disease from the homes of either parent – Europe for A. hippocastanum, the USA for A. pavia.

Infection & Development: The disease appears not to have been studied, and observations on the initiation and early development of the galls are lacking, but limited observations at very long (10- and 20-year) intervals suggest that, having attained a large size, galls enlarge very slowly indeed: Fig.64 shows galled branches in 1983; Fig.65 shows the same branches in 1993.

SECTION **3**

Fig. 63

Fig.66 was taken in 1974 while Fig. 67 shows the same gall in 1993.

Control: None known.

Remarks: The hybrid A. × carnea arose probably in Germany in about 1818. The disease is mentioned in Bean's Manual of Cultivated Trees and Shrubs, quoting an earlier writer, in 1929.

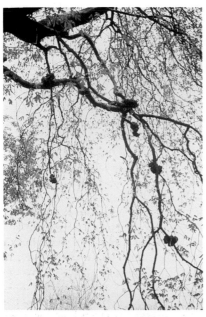

Fig. 64 *Galled branches in 1983*

Fig. 65 *The same galled branches ten years later*

Fig. 66 *Galled main stem in 1974*

Fig. 67 *The same galled stem nineteen years later*

BULGARIA CANKER   Fungus: *Bulgaria inquinans*

Fig. 68

Fig. 69

Damage Type: Stem bark killing.

Symptoms & Diagnosis: An amber-coloured gum and/or a black paste (hard when dry) exudes from the intact bark of the stem at any height (Fig.68); or sunken patches of bark up to several square feet in extent are cracked and falling away from the stem leaving irregular areas of bare wood surrounded by healthy callus tissue (Fig. 70); or a tree has died with such symptoms (Fig.69). Additional indications: Blackish brown, rubbery, top-shaped fruit bodies of the fungus, about 2 cm across × 2 cm high, are growing on the dead bark (Fig.71).

Confirmation: The bark in the vicinity of the gum or paste is dead or dying but the underlying wood is sound, though perhaps stained; fungal fruit bodies, as described above, are present on the dead bark; pieces of freshly killed bark, if wrapped in damp paper, produce a bright orange mycelium from the inner surface after a few days (Fig.72).

Caution: Gum also issues from bark killed by Phytophthora diseases (p.202, 205 and 206) or bonfires (p.122).

Fig. 70

SECTION **3**

Status: Uncertain but probably uncommon or rare.

Significance: Potentially fatal, though usually only disfiguring.

Host Trees: Quercus coccinea seems to be particularly susceptible. The literature suggests that Q. rubra is susceptible and that Fagus sylvatica may be slightly so.

Resistant Trees: The status of other oaks is not known but the lack of recorded cases of such a disease suggests that other common species of oak grown in this country are more resistant.

Infection & Development: The mode of infection has not been elucidated but must be from the rain-borne conidia or the air-borne ascospores or both (see also Remarks below). Whether unwounded bark can be infected is unknown. Once in the bark, the mycelium spreads rapidly through the phloem, killing it. This takes place during the tree's dormant season, when patches of bark approaching 2.0 m high × 0.4 m wide may be killed. Spread ceases during the growing season, when callusing begins and spores (conidia) in paste-like masses ooze from the bark fissures (Fig.68). Small parts of the callused canker margins may be reinvaded by the fungus in the second and subsequent dormant seasons, but generally the fungus fails to enlarge the initial canker, and during the growing season, after perhaps producing its rubbery, top-shaped fruit bodies (ascocarps) on the dead bark (Fig.71), dies out and is replaced by sapwood rotters such as species of Stereum. Occasionally, even large trees are girdled and killed by successive infections (Fig.69).

Control: None available: by the time gummosis is noticed, the bulk of the damage has almost certainly been done.

And if excision of active lesions is attempted, so much of the surrounding apparently healthy bark must be removed to ensure that the fungus has been eliminated that the final wound will be as large as that which the fungus might have caused.

Remarks: Bulgaria inquinans is a very common early colonizer of freshly cut oak and beech, indicating its probable ability to exist endophytically in the living tree.
The fungus was suspected of being pathogenic in 1887, in Germany, and shown to be so in 1984, in England.
This account is based largely on unpublished work by R. G. Strouts.

Fig. 71

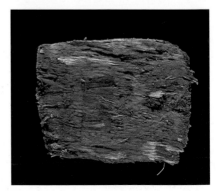

Fig. 72

CHEMICAL DAMAGE including HERBICIDE and ROAD DE-ICING SALT DAMAGE

Fig. 73

Fig. 74

Damage Type: Disruption of the tree's metabolic processes.

Symptoms & Diagnosis: Many chemicals give rise to non specific symptoms such as leaf yellowing and browning, dieback and defoliation, or cause damage which resembles that caused in other ways, e.g. leaf spotting or mottling, interveinal yellowing, bark killing. In these cases, field diagnosis depends on circumstantial evidence and a lack of any other satisfactory explanation.

The following symptoms are so distinctive that immediate consideration should be given to the possibility of chemical damage:

In conifers and broadleaves:

(i) **shoots, petioles, leaves or needles are strongly curved or twisted, cupped or otherwise distorted**; dieback or defoliation may also be evident – Typical causes: a hormone type of herbicide based on e.g. 2,4-D, dicamba, mecaprop (Figs.73-75). (But in Picea, also consider spring frost damage, p.126.)

SECTION 3

Fig. 75

Fig. 76

Fig. 77

(ii) **The youngest foliage is white or pale yellow** – Typical cause: a herbicide containing amitrole (Fig.76).

(iii) **Foliage is killed in discrete patches or swathes** and involves several branches or only part of the foliage on a single branch – Typical causes: probably a chemical which kills on contact such as de-icing (Fig.77) or blown sea salt (Fig.78); a contact herbicide such as paraquat (Fig.79); cats' urine (Fig.81).

Caution: Heat from e.g. bonfires can cause similarly distributed damage (see p.122).

In broadleaves:

(iv) **The leaves are miniaturized,** perhaps also bunched because of little or no extension shoot growth, yellowed or with a reddish tinge, and accompanied by dieback – Cause: probably a herbicide containing glyphosate (Fig.80 on hawthorn), imazapyr (especially if tinged reddish: Fig. 82 on Flowering cherry) or similar.

(v) On certain branches, most **buds have failed to flush** or, before reaching full size, the **leaves have browned and died and hang withered** on the tree (Fig.84); or during summer, the **margins and interveinal areas of leaves are brown** (Fig.85) **and the leaves are falling** prematurely. Recovery shoots may

Fig. 78

Fig. 79

Fig. 81

Fig. 82

Fig. 80

Fig. 83

SECTION 3

have developed similar symptoms. Healthy, **green leaves may be scattered on affected branches**. The bark of affected twigs is alive. Cause: probably road de-icing salt.

(vi) **Veins or** (depending on species) **interveinal leaf areas** are yellow – Typical causes: herbicides based on atrazine (Fig.83) or bromacil.

Caution: For interveinal chlorosis see also Lime induced chlorosis, p.174.

Confirmation of all cases of suspected chemical damage: Analysis of damaged plants reveals residues of the suspect chemical in sufficient quantity to have caused the damage observed.

Additional indicators: Shrubs and herbaceous plants close to the affected trees show similar symptoms; weeds (or, in a grassed area, just broadleaved weeds) are absent from the site (suggesting that herbicides have been used); the damaged plant parts or the soil smells of a chemical; the plant or soil surface is visibly coated in a dry or oily chemical; dead, wedge-shaped patches of bark are present at the base of the tree stem (suggesting the possibility of de-icing salt damage – but consider also Phytophthora root disease (p.206)).

Status: Chemical damage of one sort or another is common and widespread, particularly near buildings and roads and the sea and during the establishment phase of plantings.

Significance: Damage from herbicides, de-icing and sea salt and cats' urine, ranging from temporary disfigurement to the death of many trees, is widespread and common. Damage from other chemicals can be equally severe but most types are rare.

ACTIVITIES, CONDITIONS and EVENTS which MAY LEAD TO DAMAGE:

(a) **Herbicide applications**. The application of herbicidal sprays in windy weather, incautious application or unsuitable choice of the herbicide, or inappropriate application rate, method or timing commonly leads to damage to trees and other plants in the vicinity. Some hormone herbicides volatilize in hot, still weather, damaging foliage which overhangs the treated area (a lawn, for instance) (Fig.74). Some herbicides applied to tree stumps to prevent their regrowth may also be taken up by neighbouring trees of the same species via root grafts.

(b) **Other legitimate chemical applications**. Plants can be damaged by certain insecticides or fungicides or their diluents. Also, excessively high concentrations or unbalanced applications of artificial fertilizer can cause direct damage or nutritional disorders.

(c) **Road de-icing salt applications**. Damage arises when salt water is washed into the rooting zone of trees and other plants from roads, paved areas (Fig.84) and poorly designed storage bins or is splashed onto above-ground parts by passing traffic (Fig.77).

(d) **Wind-blown salt spray and sea floods**. Damage from wind-blown sea spray (Fig.78) is common on the coast but gales can also result in damage many miles inland. Trees inundated by sea water are likely to suffer badly.

(e) **Chemical spillages and leaks**. Almost any chemical can be damaging if present in sufficiently high concentrations around roots. The commonest problems of this nature are due to oil leaks or spillages from pipes or storage tanks. Leaks of methane (North Sea or 'natural' gas) into the soil encourages an almost explosive multiplication of certain soil bacteria which depletes the soil oxygen and asphyxiates roots.

(f) **Fumes**. In suitably high concentrations, various gaseous air pollutants can kill foliage. Visible damage is usually localized close to the point of emission. Sulphur dioxide (from coal or oil-fired furnaces) and fluorine (from brickworks and aluminium smelters) are the usual sources of such damage in this country. Visible foliar damage due to vehicle exhaust fumes can to all intents and purposes be discounted. Damage closely associated with roads, drives and car parks is most often due to deicing salt or herbicides.

Some formulations of certain hormone-type herbicides vaporize in very hot weather and in still conditions can cause considerable damage to foliage overhanging the treated area (Fig.74).

Chronic air pollution from industry and the internal combustion engine does suppress tree growth (as evidenced by comparing tree growth in ambient and filtered air) but this does not constitute damage as defined in this book so is not further considered here. Research following the popular alarm that 'acid rain' and other air pollution was causing widespread damage to trees in Europe (including these islands) and elsewhere in the 1980s failed to show that this kind of pollution was causing direct damage. Almost all the damage was caused by a variety of well known agents both biotic and abiotic.

(g) **Malicious chemical applications**. These usually involve easily procurable herbicides or common garden or household chemicals such as creosote.

(h) **Urinating cats**. Tom cats mark out their territories by spraying a phytotoxic mixture of urine and other chemicals

Fig. 85

onto convenient plants. The cypresses, which seem either to be favoured for the purpose or to be particularly sensitive to the chemicals, often bear smelly patches of dead basal foliage (Fig.81).

SPECIES SUSCEPTIBILITY: Species vary widely in their susceptibility to the various phytotoxic chemicals. This fact can be useful in diagnosis, but only where damage from road de-icing salt or sea salt can be anticipated does it help in overcoming the problem: then, the use of resistant species should be considered (Dobson 1991c, d)

Species particularly tolerant of de-icing salt contaminated soil:

Betula pendula
Gleditsia triacanthos
Paulownia tomentosa
Pinus mugo
Populus alba
Populus × canescens
Quercus species
Robinia pseudoacacia
Sophora japonica
Taxodium distichum

In a survey of salt-damaged urban trees, Gibbs (1994) found Japanese flowering cherries to be almost entirely unaffected. These common street trees do not appear in Dobson's (1991d) literature review.

Species particularly tolerant of salt spray:

Gleditsia triacanthos
Pinus mugo
Pinus thunbergii
Populus alba
Robinia pseudoacacia

Fig. 84

SECTION 3

CONTROL: **Salt**: See above and Further Reading below. **Other chemicals**: Once symptoms have appeared, no treatment is likely to be effective. Irrigation of plants contaminated with soil-acting herbicides is likely to worsen damage by increasing uptake of the chemical.

Where foliage or thin bark has been contaminated with **contact or leaf-absorbed herbicides or cat's urine**, prompt douching with water can prevent or minimize damage. Prompt removal of soil contaminated with **soil-acting herbicides** can be similarly beneficial.

Small plants in **oil-contaminated soil** can be saved by replanting in clean soil

after the oily soil has been washed from the roots. Aeration by frequent cultivation, application of a general fertilizer and liming will accelerate the biodegradation of oil and the return of the soil to normal. **Other types of damage** can only be avoided by the proper and responsible transport, storage and use of the chemicals.

Further Reading: *De-icing salt damage:* The whole subject – Dobson 1991d; Diagnosis – Dobson 1991a; Control – Dobson 1991b, Gibbs 1994; Species susceptibility – Dobson 1991c; To distinguish between salt damage and Anthracnose of London plane – Gibbs, 1983.

CHERRY BARK TORTRIX
Insect: *Enarmonia formosana*
(= Laspeyresia woeberiana)

Damage Type: Swellings on the trunk.

Symptoms & Diagnosis: Gummy exudations appear on the trunk, often associated with cracks or roughened swellings (Fig.86).

Other indicators: Light brown or rust coloured frass (excreta) mixed with fine silken threads may hang down from the site where a feeding larva is present in the bark (Fig.87).
Confirmation: Excavation with a sharp knife may reveal tunnels in the bark and cambium *but not* penetrating the sapwood.
Caution: see Remarks below and Crown gall p.135.

Status: Widely distributed and locally common in England, Wales, Scotland north to Perthshire.

Significance: Large roughened swellings result from prolonged heavy attacks. Although common this species rarely causes serious damage in Britain.

Fig. 86

Host Trees: Prunus avium, P. domestica, P. laurocerasus and occasionally P. armeniaca, P. dulcis and P. persica. Sometimes on Malus sylvestris and Pyrus communis. Larvae have also been found on Sorbus.

Life Cycle: Adult moths emerge from June to September. They fly in afternoon sunshine; at other times they can be found resting on tree trunks. They are dark in colour with yellow and white irregular markings, wingspan 14–18mm. The larvae, which are greyish-white or pinkish and with a brown head, feed from September to May and then pupate in the bark. Pale brown empty pupal cases (7–9mm long) pro-

Fig. 87

trude from the bark surface after the adults have emerged.

Control: Damage rarely justifies control measures but, if necessary, remove loose bark in March while the trees are still dormant and follow with an application of a tar oil winter wash. [Take note of current Pesticides Regulations – see p.288]

Remarks: More extensive galleries, without gummy exudation or swellings, but with circular emergence holes c. 2.5mm diameter, and protruding empty pupal cases up to 15mm long, are indicative of the much less common red belted clearwing moth Synanthedon myopaeformis. This species is found in southern England, north to Yorkshire. It occurs on Malus, Pyrus, Crataegus and Sorbus.

SECTION 3

CHERRY BLACKFLY   Insect: *Myzus cerasi*

Damage Type: Distortion and death of leaves and shoots.

Symptoms & Diagnosis: In April and May young leaves are curled and the shoots became distorted and stunted (Fig.88). By mid-summer the affected leaves are conspicuous and turn brown or black. The damage is particularly noticeable at the tips of branches where distorted leaves are bunched together.
Confirmation: From April to June there are colonies of dark brown to black aphids on the underside of the curled leaves.

Status: Common and widespread.

Significance: Persistent attacks cause distorted and weak growth, lack of bud formation, and may result in dieback. Damage is most serious on young trees or nursery stock where their future shape will be affected.

Fig. 88

Host Trees: Prunus avium, P. padus and various ornamental cherries.

Life Cycle: M. cerasi overwinters as eggs on young shoots and in bud axils. The eggs hatch before the tree blossoms in March or April. Several generations of wingless aphids occur under the protection of curled leaves until July, when winged aphids appear and migrate to herbaceous plants including Veronica (speedwells) and Galium (bedstraws). Aphids return to Prunus in the autumn when the eggs that will overwinter are laid. A few M. cerasi may remain on cherry throughout the summer.

Control: Use a tar oil winter wash during the dormant season (December–January). If this is not done then, if necessary, use a contact organophosphorus insecticide *before* the white bud stage, cutting down any flowering weeds beneath the tree prior to application to avoid killing bees. [Take note of current Pesticides Regulations – see p.288]

WARNING: Systemic insecticides are not recommended as they are damaging to some ornamental cherries.

COATINGS, DEPOSITS, EXUDATES, OUTGROWTHS and DEFORMITIES

Significance: Some coatings, deposits, exudates and outgrowths on or from various parts of a tree constitute a problem in themselves, while others are often valuable clues to the cause of a problem: fungal fruit bodies, Sooty moulds or resin bleeding, for example. Trees are, however, host to many harmless organisms. Many of these, such as mosses, liverworts, ferns, and lichens, are familiar and instantly recognizable for what they are. Others though, less familiar, can easily be misconstrued as either harmful or harmless, or may puzzle the observer, or are overlooked altogether when they might be helpful in a diagnosis. It is these which are briefly dealt with in this Section.

1. SUPERFICIAL COATINGS AND DEPOSITS
readily removed by rubbing or scraping. (These may be dried EXUDATES: see below. If in doubt, note that dried exudates will disperse or soften in water).

ON MANY SPECIES

On bark or leaves/needles
• Green, grey or orange: Algae (Fig.89). Harmless.
• Black, may also be present on ground vegetation and other surfaces: Sooty moulds (Figs.90,178). These grow on the Honeydew excreted by some harmful sucking insects, e.g. aphids (see p.144).
• Whitish appressed flecks or spots with a darker or yellow centre: Scale insects. Harmful (see p.248).
On bark
• White rings: Athelia (Fig.91), a fungal parasite of algae and lichens. Harmless to the tree. (See Rose, 1983).
• White or coloured patches: Fruit bodies of fungi (See Top rots and root rots, pp.235 and 218) or lichens (Fig.92). Some fungi are harmful, lichens are harmless.

ON FAGUS, FRAXINUS OR ULMUS BARK

- White flecks or patches (may be extensive) composed of wool-like filaments: the protective covering around scale insects (see pp.120,250 or 150). Harmful.

ON MALUS BARK

- White patches of wool-like material: the protective covering over the Woolly aphid (p.275). Harmful.

Fig. 89

Fig. 91

Fig. 90

Fig. 92

SECTION 3

2a. EXUDATES FROM BARK

FROM FAGUS, POPULUS, SALIX

Fig. 93

- Orange or yellow, gelatinous (brittle if dry), curved threads, or blobs scattered in profusion over (often) large areas of bark: on Fagus these are probably spore tendrils of the saprophytic fungus Libertella faginea (anamorph of Quaternaria quaternata) (Fig.93) or of other macroscopically similar fungi in the Family Diatrypaceae. On Populus and Salix they are spore tendrils of the saprophytic fungus Cytospora chrysosperma (anamorph of Valsa sordida). A sure indication that the bark is dead.

FROM BROADLEAVES

- Gummy, tarry or sugary: usually indicates dead, dying or injured bark. Some diseases are characterized by heavy exudates of this nature, e.g. Phytophthora bleeding canker (p.202); Beech bark disease (p.77), weeping canker of Caucasian lime (p.269). See also Drought (p.112).
- Watery: The issue of watery sap from vertical bark fissures or ribs is characteristic of Bacterial wetwood (p.74); if from the stem base, it indicates dead bark or roots. (See Fig.182).
- Paste- or yeast-like: if smelly and black, grey or orange, usually the result of colonization by yeasts and bacteria of watery or sugary exudates. See e.g. Bacterial wetwood (p.74); if odourless, black, orange or yellow and in blobs or filaments, probably fungal spore masses and an indication of dead bark. See e.g. Bulgaria canker (p.91) and Fagus etc. above.

FROM CONIFERS

- Resin: Burst resin blisters and small runs of resin are normal on many species, e.g. Abies, Pinus, Pseudotsuga. Otherwise they are an indication of dead, dying or injured bark or, if from the lower stem, of dying roots (see e.g. Dendroctonus (p.138), Honey fungus (p.145), Coryneum canker (p107); Blister rust of white pine (p.83)).
- Watery fluid from e.g. Taxus: Rainwater which has accumulated in a multiple fork may slowly seep out through fissures where several stems have fused together imperfectly to form one bole.

2b. EXUDATES FROM WOOD

FROM BROADLEAVES

- Watery, from pruning wounds: some species (Acacia, Acer, Betula, Carpinus, Gleditsia, Gymnocladus, Juglans, Laburnum, Magnolia, Populus, Pterocarya, Robinia, Sophora) bleed profusely but without evident harm from wounds made in late winter or spring. Avoid this by pruning these in late summer or autumn.

3.   OUTGROWTHS

These are structured entities. Their removal may require force and may damage the substrate.

ON ANY SPECIES

From bark of stem, branches or exposed roots: (see also Galls, p.132)
- Coloured variously: bracket-, shelf-, hoof- or toadstool-like, or roundish lumps; woody, leathery or fleshy; or thick, brittle encrustations: fruit bodies of wood-rotting or bark-killing fungi (See Figs.71,183,316-352,374-5,382-3).

From branches:

- Abnormally dense, large or small, often spherical clusters of twigs in an otherwise normal tree crown: witches' brooms (see p. 270); or (on broadleaves only), if the twigs bear simple, evergreen leaves in pairs differing from those of the tree itself: mistletoe, Viscum album: see Fig. 94 and 'Deformities' below.

From foliage, flowers and fruit:
see Galls, p.132.

From roots, root collar or stems:
- Spherical, whitish, soft pea-like galls, or larger, bark-coloured, woody galls: crown gall (p. 132).

Fig. 94

FROM ACER, LIQUIDAMBAR AND ULMUS TWIGS

- Bark-coloured, irregular ribs, wings and pegs running along the twigs ('winged cork'): Normal corky outgrowths of Ulmus procera, Liquidambar styraciflua (Fig.95), Acer campestre and a few others.

Fig. 95

SECTION 3

FROM FAGUS BARK

Fig. 96

Fig. 97

- Black, hard, slightly raised sheets, or patches and spots arranged in horizontal lines on the lower stem: The fungus Ascodichaena rugosa (Fig.96). Harmless.
- Bark-coloured spheroid lumps, small or large: Sphaeroblasts (Fig.97), i.e. woody, bark-covered structures, being an abnormal development of a bud which has produced annually a woody sheath without ever producing a shoot. Harmless.

4. DEFORMITIES

Many pest and disease problems include various deformities among their symptoms (e.g. Pine shoot moth, p.211; Pocket plums, p.213; Taphrina leaf diseases, p.197) but these should be identifiable by referring to the host tree involved in Section 2. The following occur on a wide range of trees.

ON ANY SPECIES (LEAVES, SHOOTS AND TWIGS)

- Curved markedly; petioles or needles may also be affected; leaves may be cupped: see herbicide damage, p.93.
- Flattened markedly, perhaps appearing to consist of several shoots joined together longitudinally, often curved: this condition is known as fasciation (Fig.99). It is the result of the normally dome-shaped growing point at the shoot tip changing into a narrow row of growing points. No causes have been identified. A usually transitory curiosity.

Fig. 98

Fig. 99

ON BROADLEAVES (BRANCHES)

- Swellings. The parasitic common mistletoe (Viscum album) sometimes, instead of producing the familiar balls of leafy twigs, produces few leaves from grossly swollen and deformed portions of infected branches (Fig.98).

CONIFER SPINNING MITE   Mite: *Oligonychus ununguis*

Damage Type: Discoloration, death and fall of needles.

Symptoms & Diagnosis: The needles have a yellow speckled appearance at first but later the foliage turns to a bronze colour (Figs.100 and Fig.102). This change develops from the centre of the tree's crown affecting needles on the current shoots last; it is most noticeable in late summer or autumn. Loss of severely damaged needles may occur.
Confirmation: Close examination will reveal tiny orange-brown to grey spider mites (0.2–0.5mm) and orange-brown spherical eggs (Fig.101) on the shoots. Hatched eggs are colourless and may persist on the stems for some time after an attack has finished; they are most easily seen on the undersides of affected branches. The mites spin silk but this is only obviously visible when very large numbers are present.

Status: Common; occurs throughout Britain.

Significance: Most often damaging where young trees are growing in dry conditions. A serious problem on Christmas trees in some years and will also cause severe damage to dwarf cultivars of spruces.

Host Trees: Many Picea species are vulnerable, particularly young trees, but damage is most often reported on P. abies, a reflection of this spruce's importance to the Christmas tree trade rather than an indication of host preference by O. ununguis. This mite also sometimes causes damage on many other conifer species.

Life Cycle: Overwintering eggs hatch in April to mid May and the larvae, which have three pairs of legs, develop through two nymphal stages before becoming adults with eight legs. Eggs are laid by the adult females on the

Fig. 100

Fig. 101

SECTION **3**

needles and stems throughout the summer, leading to rapid population increases under favourable weather conditions. O. ununguis overwinter as eggs on the current shoots, mainly clustered on the bark close to the terminal buds.

Control: : In late May or as soon as the eggs have hatched, apply a permitted acaricide active against adult mites. Applications later in the season will require an acaricide which is active against both mites and eggs. [Take note of current Pesticides Regulations – see p.288]

Further Reading: Carter & Winter (1998).

Fig. 102

CORAL SPOT

Fungus: *Nectria cinnabarina*
Anamorph: *Tubercularia vulgaris*

Damage Type: Bark killing.

Symptoms & Diagnosis: Scattered twigs have failed to flush, or shoots (perhaps branchlets) have wilted soon after flushing (Fig.103).
Additional indicators: Affected plants have recently been subjected to a stress such as transplanting, drought, or waterlogging. Dead bark patches are centred on injuries from such things as chafing stakes, tree ties, or pruning.
Confirmation: Dead twigs, or patches of dead bark surrounding wounds or the base of wilted shoots, are covered in globose, bright pinkish pimples 0.5–1.5 mm across, darkening with age (The fruit bodies of the fungus Fig.104). Dead bark patches are centred on injuries.
Caution: If many Coral spot infected twigs are present or large areas of bark are colonized, always consider the possibility that the tree is or has been suffering from a more generally debilitating condition (e.g. a root or vascular wilt disease, adverse growing

Fig. 103

conditions, chemical damage) or from a different bark killing condition (e.g. fire damage, or a bacterial bark disease) and that the Coral spot has merely invaded dead or dying tissues.

Status: Common and widespread.

Significance: Because the fungus is such a common colonizer of bark killed by other agents, its true pathogenic status on trees is unclear; its presence is often a sign that the tree is or has been under severe stress or that the colonized tissues have suffered prior injury.

Host Trees: The fungus has been reported on a very wide range of broadleaves, most often on Acer, Aesculus, Carpinus, Fagus, Juglans, Tilia, Ulmus, and occasionally on various conifers.

Infection & Development: Spores are produced from the pink 'Coral spot' fruit bodies which grow out of the killed bark, and are dispersed in rain. Infection of the shoots or twigs of very susceptible trees, or trees under water stress, probably takes place via natural openings such as lenticels or bud scale scars. Mechanical wounds caused by pruning or abrasion, and winter-cold damaged bark, or spring-frost killed flowers (in magnolia, for example) are also common infection courts. Spread in stressed or unhealthy trees can occur at any time; in otherwise healthy trees, the fungus kills bark and sapwood during winter dormancy but its spread is halted once the tree resumes growth in spring.

Control: *Therapy:* Identify and if pos-

sible rectify any cultural or underlying disease or other problem (e.g. improve irrigation, soil conditions, weed control). The excision of dead bark or removal of dead twigs is doubtfully of any but cosmetic benefit.
Prevention: Provide the best possible growing conditions. Avoid bark injuries, especially to recently transplanted trees.

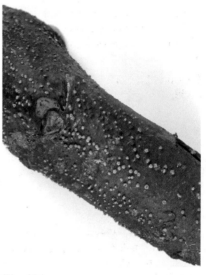

Fig. 104

SECTION **3**

CORYNEUM CANKER Fungus: *Seiridium cardinale*
(= Coryneum cardinale)

Damage Type: Perennial bark killing.

Symptoms & Diagnosis: Scattered patches of faded, yellow or brown foliage are evident in the crown, and leafless twigs and branches may be present; the whole top of the tree may be yellow or dead (Figs.105,106).
Confirmation: Extensive areas of dead bark, often girdling and encrusted with and exuding resin (Fig.107), are present on the branches or stem bearing affected foliage. On the bark that has been dead for some time can be found patches of pin-head sized, black pimples

(fruit bodies) or tiny crater-like cavities (where fruit bodies have disintegrated).

Status: Common in England and Wales; by 1999 reported only once from Scotland.

Significance: A slow-spreading but very disfiguring and ultimately often fatal disease.

Host Trees: Confined to Cupressaceae (Cypress family). Cupressus macrocarpa (Fig.105) and C. sempervirens are very susceptible. Also on × Cupressocyparis leylandii (Fig.106) on which its full potential is uncertain: severe damage does occur but reported cases are still very few in relation to the enormous number of Leyland cypresses there are. A few cases of appreciable damage have also been noted on Thuja plicata in this country.

Resistant Trees: Chamaecyparis lawsoniana is probably immune. Ch. nootkatensis, Cupressus arizonica, C. glabra, Thuja occidentalis, and Juniperus virginiana are highly resistant. Calocedrus decurrens, J. chinensis and various other Cupressaceae grown here are reported to suffer occasional slight damage abroad.

Infection & Development: Spores are spread in rainwater and possibly by insects. Infection is via small natural bark wounds such as cracks in branch crotches. The fungus spreads through the bark, killing it, girdling twigs within one season and larger branches over a number of years. It eventually spreads down into the stem. The tree dies either if it is girdled below the live crown, or from multiple branch cankers. Fruit bodies develop from infected bark.

Control: The removal of affected branches by cutting well below dead bark will retard disease development and may be worth undertaking on recently symptomatic trees. In a tree which has been infected for any length of time, this procedure is likely to be unacceptably disfiguring, however, and many infections are likely to remain undetected. Badly diseased trees are therefore best replaced with resistant species. New plantings with the more susceptible species are ill advised near infected trees.

Remarks: The disease was first noted in the USA, in 1928. First reported in Europe in 1944, in England in 1969 and in Scotland in 1996. The first reported case in England on Leyland cypress was in 1982.

Further Reading: Strouts 1990.

Fig. 105

Fig. 106

Fig. 107

CYPRESS APHID   Insect: *Cinara cupressi*

Damage Type: Death of foliage and branches.

Symptoms & Diagnosis: Patches of yellow or straw-coloured current and older foliage appear in late spring and summer, particularly at the base of the tree or hedge; these later turn brown and appear dried-up (Fig.108). Dieback of the lower branches may spread upwards, and in hedges sideways, to adjacent plants.

Other indicators: Sooty moulds and honeydew (see p.144) are present on the stems and foliage. In late summer wasps may be attracted by the honeydew.

Confirmation: Look for aphids on thinner barked stems and foliage from May to November. These aphids are fairly large (1.8–3.9mm long), pear-shaped and greyish to yellow-brown with two dark stripes on their back which diverge from the head. There are prominent black siphuncular cones (cornicles) on the abdomen which are connected by a transverse black band (Fig.109).

Status: Widespread and common in some years. Most frequent in southern and eastern England, but damage has been reported as far north as central Scotland.

Significance: Causes thinning of the crown and loss of lower branches on specimen trees, hedges and sight screens. If screens are attacked, a serious loss of the screening effect may occur as there is little if any regrowth from old wood on the Cupressaceae.

Host Trees: Probably all cultivars of × Cupressocyparis leylandii can be affected, but the golden 'Castlewellan' seems particularly susceptible. C. cupressi also occurs on Chamaecyparis lawsoniana, Cupressus macrocarpa and occasionally dwarf cultivars of Thuja occidentalis. C. lawsoniana is rarely affected as badly as × C. leylandii.

Life Cycle: Incompletely known. Colonies of viviparae (the aphid stage that reproduces by live birth) are known to occur from late January until the end of November in south-west England. Winged forms appear from June to August. Although egg laying stages (oviparae) are known to occur in October, it appears that viviparae can overwinter under mild conditions and this may be the usual behaviour in southern Britain.

Control: Spray the affected plants immediately symptoms are noticed with an insecticide approved for the control of aphids on trees or hardy ornamentals. Good results have been achieved

SECTION 3

Fig. 108

Fig. 109

using pirimicarb. [Take note of current Pesticides Regulations – see p.288] Dieback may continue for a while after the aphids are killed.

Remarks: A southern European species that was first recorded in Britain in the late nineteenth century.

Damage to Cupressus macrocarpa was noted in southern England between 1920 and 1940. Damage to × Cupressocyparis was first noticed in the 1980s and became especially widespread and severe in 1988 and 1989.

Further Reading: Winter (1989).

CYTOSPORA DIEBACK OF CHERRY LAUREL
Fungus: *Cytospora laurocerasi*

Damage Type: Bark killing.

Symptoms & Diagnosis: In spring or early summer, all the leaves on scattered branches or shoots have yellowed, wilted or hang dead on the plant (Fig.110). Symptoms may have recurred for several years.
Additional indicators: On a branch not entirely dead, a patch of dead bark, with or without a brownish gummy deposit, may encircle the branch at its junction with a smaller dead side branch. At this point, and on other large but live branches, a slightly sunken, rough-barked, gummy canker may be present (Fig.111).
Confirmation: If a piece of branch incorporating the dead/live junction is kept wrapped for a week or so in damp newspaper, the amber or yellowish spore tendrils of the fungus may develop (Fig.112).
Caution: Rapid and severe dieback within a year of pruning large branches is more likely to be due to Silver leaf (p.252).

Status: Common in the southern half of England. Its status in the rest of the country is not known.

Significance: A very disfiguring disease, though plants quickly recover if infections are cut out. If untreated, it seems that the disease can sometimes eventually kill sizeable plants.

Fig. 110

Host Trees: To date, recorded only on Prunus laurocerasus.

Infection & Development: This has not yet been fully elucidated. The fungus perennates in infected bark and has been found causing small, brown leaf spots which later become holes (cf the 'shothole' symptom caused by Bacterial canker of cherry (p.70)). Tendrils of spores are extruded from infected tissues in wet weather (Fig.112) and are probably dispersed in rain. Pruning wounds can be infected artificially and are often associated with natural infections. Just as often, however, infection appears to have taken place through buds or shoots without obvious wounding. Limited evidence suggests that infection is most likely in autumn or

Fig. 111

Fig. 112

spring but most bark killing occurs during the tree's dormant period. An amber, gummy liquid exudes from the dying bark, sometimes copiously (Fig.111). Shoots and twigs die as the fungus spreads down them. If the fungus then reaches a larger branch a perennating canker is formed which slowly spreads round the branch and may eventually girdle and kill it.

Control: *Therapy:* Cut infected branches off well below the limit of dead bark, back to a healthy side branch or bud. Infection of the pruning wounds by Cytospora or Silver leaf (p.252) is unlikely if cutting is done in June, July or August. At other times of year, it is as well to apply a fungicidal wound paint. [Take note of current Pesticides Regulations – see p.288]

Prevention: The indications are that trimming hedges by shearing one-year-old shoots does not invite infection. However, if older wood is to be cut, it is sensible to do this in summer or to use a wound paint, as mentioned above.

Remarks: No published account has been traced for such a disease on Cherry laurel but cases have been recorded by the Forestry Commission since 1974. The assumption that Cytospora is responsible rests on one unpublished but successful inoculation experiment by R. G. Strouts, the common occurrence of the fungus in dead and dying bark, and the repeated failure to find other fungal or bacterial pathogens.

DROUGHT DAMAGE   Cause: *Lack of available soil water; faster water loss from leaves than uptake from soil*

Fig. 113

Fig. 114

Damage Type: Growth reduction, defoliation, dieback; death.

Symptoms & Diagnosis: (for distinctive symptoms peculiar to certain species, see below). The damage that is evident varies according to tree species but the following is common to most: *Early effects:* Foliage wilts in the day (Fig.114) but recovers at night; the older needles or leaves are faded, yellow or brown and falling (Figs.113,114,118). *As drought intensifies:* The younger leaves of broadleaved trees are faded or exhibit brown blotches, or the edges and interveinal areas are yellow or brown; leaves and younger needles are dying and falling. *If the drought continues:* Leaves and needles of all ages have wilted and died and are falling (or whole shoots have died and bear brown and shrivelled leaves); twigs and branches have died back; gummy exudations are present here and there on the stems of broadleaves (Fig.115); longitudinal splits are evident in some tree stems (Fig.116); some trees, especially small or newly planted ones, are dead. *In following years:* Crowns appear thin due to twig death, small leaf size and poor extension growth or (in evergreens, especially

Fig. 115

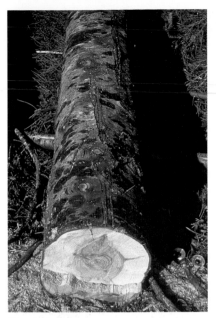

Fig. 116

Fig. 117

SECTION **3**

conifers) to the lack of older needles or leaves; patches of dead bark on stems are cracking and falling away to reveal callus tissue beneath; old stem cracks appear as longitudinal bark ribs or scars (Fig. 116 on Abies).

Additional indicators: Long periods of hot, dry and perhaps windy weather have preceded the onset of symptoms. The soil is sandy or gravelly, or shallow over rock (not chalk).

Caution: The appearance of drought symptoms sometimes indicates that a tree's roots are already damaged by, for example, disease or compaction. See also Beech bark disease (p.77) and Exudates (p.100).

Additional characteristic responses of certain species:

Acer pseudoplatanus: Small, diamond-shaped patches of bark die on stems, revealed later when callusing forces dead bark off (Fig.117).

Fig. 118

Betula: From a distance, the scattered, yellow leaves give the tree's crown a speckled appearance.

Chamaecyparis: The normally green foliage towards the inside of the tree yellows, browns and falls while the young foliage at the branch extremities remains green. If drought seems improbable, the possibility that such symptoms are due to magnesium deficiency should be explored.

× *Cupressocyparis leylandii:* Like Chamaecyparis above. In the case of Clone 20, scattered small fronds assume a brownish colour on one surface but a drab, glaucous grey on the other before dying and falling (Fig.119).

Fig. 120

Fig. 119

bodies of only one fungal species: often Biscogniauxia nummularia (Fig. 120) (=Hypoxylon nummularium).

Larix: Tops of trees die.

Liriodendron: Leaves develop innumerable tiny, brown, interveinal spots before falling (Fig.121).

Fig. 121

Fagus: Lightning-type strips of dead bark, often involving a damaged root or branch, a branch stub or pruning wound, or stem-base infection by Ustulina deusta (q.v.) may develop during very dry or exceptionally hot years. Apparently, these conditions allow fungi lying widespread but normally dormant in the sapwood (endophytes), to develop rapidly and to invade and rot the sapwood and kill the bark. If evidence of lightning damage is absent these drought-induced 'beech strip cankers' can be suspected if the dead bark bears along its whole length fruit

Quercus: Whole small twigs with dead leaves attached are shed (cladoptosis, p.50).

Taxus: Second year needles and older yellow from their bases upwards before falling (Fig.113).

Thuja: Whole fronds yellow and die but remain attached, scattered over crown ('flagging'). Tops of trees die.

Tilia: All leaves on scattered twigs yellow and die but remain attached, apparently because twigs are girdled at base. Whether a bark infection is involved is unclear.

Significance: Drought reduces the resistance of trees to many pests and diseases as well as causing damage in its own right.

Resistant and Susceptible Trees: Most trees grown for amenity in Britain tolerate or recover from a degree of intermittent droughting, though after varying degrees of distress.

Species which rarely show symptoms:

Eucalyptus

Pinus spp

Pseudotsuga menziesii

Quercus spp

Robinia pseudoacacia

Species which soon show symptoms:

Abies spp (stem cracks)

Acer pseudoplatanus

Alnus spp

Betula spp

Chamaecyparis spp

× Cupressocyparis leylandii Clone 20

Fagus sylvatica

Fraxinus excelsior

Larix spp

Picea spp

Taxus baccata

Thuja spp

Tilia spp

The tree's need for water: Water is essential for photosynthesis and other metabolic processes, transport of metabolites and minerals within the tree, and for maintaining turgor in soft tissues. A prolonged shortage is therefore potentially fatal. The progressive death and shedding of leaves reduces water loss from the plant as a whole, thus enabling it to prolong its survival on the diminishing supply of water.

Control: *Therapy:* Irrigation can reverse early symptoms, particularly in small, young trees, but care must be taken to avoid waterlogging.

Prevention: Choose species appropriate for the site and climate; irrigation can prevent damage from worsening. For newly planted trees, the essential drought preventative measure is good weed control.

Remarks: See also 'Physiological drought' under Winter cold (p.129).

Further Reading: Binns 1980; Davies 1987.

SECTION **3**

DUTCH ELM DISEASE

Fungus: *Ophiostoma novo-ulmi*
Also: *Ophiostoma ulmi*
(= Ceratocystis ulmi)

Fig. 122

Fig. 123

Damage Type: Disruption of the tree's metabolism and water transporting system.

Symptoms & Diagnosis: *Syndrome A:* At any time from early summer to early autumn, patches of leaves in the crown, each patch arising from an individual branch, are wilted or yellowed, or browned and falling (Figs.122, 123); some shoots among the affected foliage are dead and may be hooked over at their ends like little shepherds' crooks (Fig.123). *Syndrome B:* A tree (unhealthy or healthy the previous summer) fails to flush or flushes and quickly dies; or during summer, the whole of a previously healthy tree is wilting, yellowing or defoliating, or is leafless; adjacent trees (notably U. procera) die in succession over one or more years (Fig.124).
Confirmation: If a symptomatic branch is cut across, a ring of dark brown marks is evident in the wood; cut lengthwise, the wood shows longitudinal dark brown streaks (Fig.125). In fresh infections, the stain may appear only in the annual ring immediately beneath the bark. Symptoms usually spread until the whole tree is dead – a process which may take only three or four weeks or as much as two years.
Caution: The brown streaking is not always present in all parts of affected branches. Verticillium wilt (relatively rare in elm) causes similar symptoms (see p.261).

Status: By far the commonest fatal disease of elm. In the 1960s, a new, highly pathogenic or 'aggressive' strain of the fungus (now named Ophiostoma novo-ulmi) was introduced into England from North America. By 1992 this had spread throughout England and Wales

Fig. 124

Fig. 125

Fig. 126

and almost as far north as Inverness in Scotland. The original or 'non-aggressive' strain of the fungus (now renamed Ophiostoma ulmi), which causes a far less damaging disease, has been largely replaced by the 'aggressive' strain, O. novo-ulmi.

Significance: O. novo-ulmi is the cause of the most devastating tree disease this country has experienced in recorded history and is still a very serious threat to our remaining elms. Millions of trees were killed during the 1970's and the new elm populations that are growing up in the aftermath of the epidemic from the surviving root systems are now being attacked. O.ulmi causes a much less serious disease from which trees often recover.

Host Species: The disease occurs only on Ulmus and Zelkova. Notably susceptible to the aggressive O. novo-ulmi are U. procera (English elm), U. hollandica (Dutch elm), and U. carpinifolia var. sarniensis (Wheatley or Jersey elm). U. glabra (Wych elm) and some varieties of U. carpinifolia (Smooth leaved elm) become infected less readily and die more slowly. Zelkova, although sometimes infected, is rarely killed. See also Control below.

Infection & Development: *Syndrome A* (Fig.122) (as described under Symptoms above): The fungus grows in the bark of infected, dying trees where it sporulates in the breeding galleries of Elm bark beetles (Scolytus species). In spring and early summer, young beetles emerge and, carrying spores on their bodies, fly into the crowns of other elm trees to feed on the young, sappy bark in twig crotches. Spores become lodged in the feeding grooves (Fig.126) where they germinate. The fungus grows into the water-filled xylem vessels and there produces yeast-like spores.
Syndrome B (Fig.124) (as described under Symptoms above): If roots of healthy elms are naturally grafted to

SECTION **3**

those of infected trees, or if (as is commonly the case with U. procera) trees in a woodland or hedgerow originated as suckers arising from a common root system, the fungus can move from tree to tree via xylem vessels in the roots.

As the fungus moves through the xylem vessels it produces toxic chemicals (phytotoxins) which disrupt the tree's metabolism. Plugs of phenolic materials – the brown marks in the wood – are formed in the vessels. The tree, dying from the complex effects of the infection, which are not yet fully understood, is open to invasion by Elm bark beetles: in summer and autumn, the females lay their eggs in galleries (tunnels) which they bore between the bark and the wood, often at the same time introducing the fungus to the bark as spores carried on their bodies. The larvae which hatch from the eggs feed by extending these galleries, then pupate either in the bark or in the outer sapwood. The beetles develop from the pupae and then eat their way through the bark of stems and larger limbs, peppering them with small, circular emergence holes.

Control: *Therapy:* Infections in specimen trees can often be arrested by means of fungicidal injections. At the time of writing, an 'Off-label' Approval for the use of a proprietary brand of thiabendazole was extant. Names of contractors who provide an injection service should be available from the distributors of this fungicide: see the current edition of the UK Pesticide Guide published by the British Crop Protection Council and CABI, or contact one of the establishments listed in Appendix 2, p.306.

For injections to succeed:

1. The tree should be at least 30 metres from other elms.

2. Symptoms should not have appeared until after 1 July.

3. No more than 5% of the crown should show symptoms.

4. Symptoms should be at least 3 metres from the stem.

5. One month after injection, the diseased branch should be removed, or the tree should be pollarded in the following winter.

6. The tree should be in an area where the disease is not prevalent.

Injection is likely to fail if:

1. Symptoms are the result of infection in the previous year (usually the case if symptoms appear before 1 July).

2. More than 10% of the crown shows symptoms.

3. The tree has been pollarded recently.

4. Symptoms are evident closer than 3 metres from the stem.

5. Infection has been via the roots.

The treatment must be given at the earliest opportunity after symptoms appear in accordance with the directions supplied.

Prevention: The prevention of the disease requires the implementation and determined continuation of a carefully planned strategy over a large area involving the co-operation of all local authorities, land owners and private householders with elm trees. The plan requires the prompt reporting and felling of diseased trees and the rapid and safe disposal of potential beetle breeding material. Legislation exists to facilitate such measures in certain parts of the country (see Section 4, p.283).

Disease-resistant trees: Several elm cultivars relatively resistant to the aggressive O. novo-ulmi have been bred in Holland and the USA. Some are on the market, the qualities of others are still being assessed and the breeding programme is continuing. Whether any of these will have enough of the attributes of any of our British elms to make their planting worthwhile, time alone will tell. Meanwhile, one approach to the problem of Dutch elm disease is to replace diseased trees with different species altogether.

Remarks: The disease was first recorded in France in 1918, had reached this country by 1927, and North America by 1930. The cause (Ophiostoma ulmi) was elucidated in Holland, hence Dutch elm disease. In the early 1960s, O. novo-ulmi reached England from Canada (see Status above). In Continental Europe the picture is further complicated by the presence of two races of the aggressive O. novo-ulmi.

Further Reading: The disease: Gibbs, Brasier and Webber, 1994.
Regrowth from rootstocks of killed English elms: Greig, 1993.
Potential for controlling O. novo-ulmi by fungal viruses: Sutherland and Brasier, 1994.
Replacing killed elms with other genera: Mitchell, 1973.
Breeding resistant elms: Burdekin and Rushforth, revised Greig, 1988.

ERIOPHYID NEEDLE MITE   Mite: *Nalepella haarlovi*

Fig. 127

Fig. 128

Damage Type: Discoloration and fall of needles.

Symptoms & Diagnosis: The foliage appears greyish at first but later turns reddish-brown (Fig.127). The damage usually begins in the interior of the crown, where it is most severe, and then spreads outwards affecting the current year's needles in the most severe cases.
Confirmation: Close examination (×10 lens) may reveal tiny pale orange or pink spherical eggs on the needles, together with sausage-shaped eriophyid mites, a little darker in colour than and 2 or 3 times the size of the eggs (Fig.128).
Caution: Similar symptoms are caused by the conifer spinning mite but its eggs are much larger, easily visible on the bark at ×10 (see p.105), and it also spins fine silk threads.

Status: Reported from Hampshire, Surrey, Derbyshire and in central Scotland.

Significance: Its full potential is not yet known but it does cause severe needle damage leading to defoliation, which may be related to dry conditions.

Host Trees: Picea abies and P. sitchensis; other spruces are probably susceptible.

Life Cycle: Incompletely known. Occurs mainly as eggs during the winter, but live mites have been seen in December and March.

Control: Use permitted acaricide active against both adults and eggs, as soon as damage is noticed. [Take note of current Pesticides Regulations – see p.288]

Remarks: So far only reported on small trees (<3m) and nursery transplants but has the potential to become more widespread in the future and affect ornamental trees.

Further Reading: Carter & Winter (1998).

FELTED BEECH SCALE (COCCUS)   Insect: *Cryptococcus fagisuga*

Fig. 129

Damage Type: Malformation and fissuring of bark.

Symptoms & Diagnosis: The trunk appears white, due either to scattered white spots or a more complete 'white-wash' (Fig.129). The bark may be dimpled, pitted or fissured (Fig.130).

Confirmation: Close examination shows the white substance to be a waxy wool that has a felt-like appearance when covering large areas of bark. This wax is secreted by tiny yellow larvae and adults of C. fagisuga which can be found mingled with it. It builds up over a period of years; older infestations become blackened by Sooty moulds growing on the honeydew produced by the insect (see also p.144).

Status: Widespread, probably wherever the host occurs. The most severe infestations in woodlands have occurred in southern England.

Significance: Often only single trees, or small groups, within a stand are affected. Individual trees show resistance to C. fagisuga and this is dramatically illustrated sometimes on grafted beech when only the stock or the scion is infested, the other half of the graft apparently being unsuitable for an infestation to develop (Fig.131). Attack by C. fagisuga is the first stage in the development of Beech bark disease (p.77) though not all infestations cause the disease to develop. Often, by the time the disease is evident, only small patches or flecks of waxy wool are present on the diseased trees.

Host Trees: Fagus sylvatica. It has been suggested that the weeping beech (F. sylvatica var. pendula) and the copper beech (var. purpurea) are more resistant than the ordinary form. Other beeches are known to be susceptible, e.g. F. grandiflora in the north-eastern USA.

Life Cycle: Adult beech scales are pale yellow, 1mm long, sedentary and present on the tree from May/June to September. They have no legs and are attached to the bark by their mouthparts only. They lay eggs in wax 'ovisacs' from June to September. The active 'crawlers' hatch from the eggs between July and November but mainly in September. Most disperse a short distance on the parent tree but a few may be carried on air currents away from the source beech. Crawlers settle down to feed and their bodies soon become so distended that they cannot move again.

Control: C. fagisuga infestations can be reduced greatly by scrubbing the affected trunk with water containing a mild detergent.

Further Reading: Lonsdale & Wainhouse (1987).

Fig. 130

Fig. 131

SECTION 3

FIRE DAMAGE   Cause: *Bonfires*

Fig. 132

Fig. 133

Damage Type: Foliage and bark killing.

Symptoms & Diagnosis: Foliage damage: A discrete patch or vertical strip of dead foliage or leafless twigs extends upwards on one side of the tree from the lowest branch (Fig.132). Bark damage: Cracked, loose, blackened or weeping areas of bark, or patches of bark peppered with insect exit holes, are evident on one side of the lower stem of the tree and perhaps also on the underside of the branch or branches immediately above.

Additional indicators: The distribution of damaged foliage will be consistent with the distribution of heat (perhaps overlapping onto branches of an adjacent tree, for example, and probably not confined to any one branch) (Figs.132, 133, and Fig.134 which shows heat damage sustained by elms from the use of road surface burning machinery prior to resurfacing). Usually, any callus tissue at the edge of stem lesions will be alive (unless a bonfire has been set on the same site more than once).

Confirmation: When probed, the affected areas of stem bark are found to involve one large, discrete patch of dead bark. This may or may not extend to ground level but below ground the bark is alive. The dead bark may or may not be charred. Charred bonfire remains may be found on the ground or concealed in vegetation facing and a short distance from the dead stem bark.

In woodland or in groups of trees, several trees in a cluster may exhibit similar stem damage, all of it facing inwards towards the site of the bonfire (Fig.135).

Caution: Large above-ground stem lesions can be due to some bark diseases e.g. Bulgaria inquinans (p.91), Phytophthora bleeding canker (p.202) and to lightning.

Status: Common on building sites and in gardens, occasional elsewhere.

Significance: Initially only disfiguring but may lead to sapwood decay and, depending on the tree species, ultimately render the tree unsafe.

Tree Species Affected: The stem damage is commonest on thin-barked trees such as beech.

Control: *Therapy:* Scorched branches may put out fresh leaves. Dead parts can be pruned out for appearances' sake. No treatment is available for killed bark (see Section 5, Decay and Safety). *Prevention:* Do not build bonfires near trees!

Remarks: This kind of bark damage is often puzzling as it may show up only after some years of subsequent callus growth, and perhaps occurred long before the present occupant of the property moved in. Dating the injury from the callus growth may link it neatly to earlier building or other operations.

Fig. 135

Fig. 134

SECTION 3

FIREBLIGHT   Bacterium: *Erwinia amylovora*

Fig. 137

Fig. 136

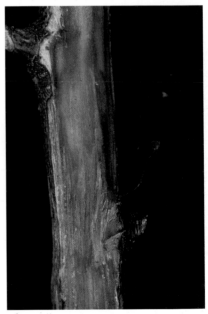

Fig. 138

Damage Type: Flower, shoot, twig, larger branch and whole tree killing.

Symptoms & Diagnosis: A few or many flowers, whole flower trusses and/or shoot tips have wilted and turned brown or black, perhaps in numbers so large that the tree looks as if it has been swept by fire (Fig.136); cf Fig.133; the shrivelled shoots are hooked over at the ends. The dead flowers and leaves slowly disintegrate and fall over a long period; twigs, large branches or whole trees may die; severe damage may develop within one growing season or over several.

Confirmation: Dead bark is present on affected parts, perhaps visible as sunken, darkened or cracked areas (Fig.137), or confined to flowering spurs; under or close to this the cambium is stained a reddish brown (Fig.138) discoloured streaks perhaps extending some distance back along the branch or down the stem beneath live bark. Drops of whitish mucilage (bacterial ooze), later darkening, may exude from dead tissues in spring and summer (Fig.139). *Additional indicators:* Extensive new damage has occurred a week or two after wet, warm and perhaps very windy weather (see Infection &

Development below). Diseased Pyracantha and Cotoneaster and other susceptible woody plants are present.
Caution: Death of scattered twigs may be due to Nectria canker (p.185). See also Resistant Trees, below.

Status: Widespread in England and Wales except in the northernmost English counties. Also in scattered localities further north and still spreading.

Significance: Although potentially fatal, fireblight is a much less alarming disease than commonly supposed: many ornamental trees suffer a degree of dieback, then recover; few are especially susceptible; outbreaks are sporadic and severe outbreaks are often localized; if caught early, surgery can limit the damage. It is more of a problem in nurseries and on susceptible shrubs as the small plants are readily killed, and in orchards, where susceptible trees may be killed and the cropping of more resistant trees seriously reduced.

Host Trees: Recorded on many trees and shrubs in the sub-family Maloideae (plants with apple-like fruits) of the family Rosaceae but in this country causes significant damage from time to time on only a few amenity tree species, most often Sorbus aria and Crataegus monogyna. The relative susceptibility of C. laevigata (= C. oxyacantha) is unclear. S. aucuparia appears to be considerably less liable to severe damage than S. aria.
In the literature, the following Crab apples and ornamental pears are stated to be susceptible: Malus hupehensis, M. 'Red Jade', M. sargentii 'Rosea', M. tschonoskii, M. 'Wintergold' and M. × zumi; Pyrus calleryana 'Autumn Blaze' and P.c. 'Redspire'. Among shrubs, many Cotoneasters and Pyracanthas are very susceptible.

Fig. 139

Resistant Trees: No Prunus species nor any other plant not in the sub-family Pomoideae is attacked.
Sorbus intermedia can be regarded as immune, and in the literature, Malus 'Liset', M. 'Professor Sprenger' and Pyrus calleryana 'Bradford' are said to be very resistant.

Infection & Development: The bacteria overwinter in bark at canker margins (Fig.137). In spring, they multiply and ooze from the bark in mucilaginous drops (Fig.139) and are distributed by insects, birds, rain, on pruning tools and, when dried, in the wind. Infection can take place via any injury or natural opening (pruning wounds, wind- or hail-damaged leaves, insect damage, stomata, lenticels etc.), but probably most often through undamaged floral parts, from bacteria carried there by pollinating insects. The bacteria quickly multiply and spread in and kill the phloem and cambium. If infection spreads down a branch into a larger one

SECTION 3

or into the stem, this may be girdled and all parts above the girdle will die, thus an infection low down a main limb or on the stem can quickly kill most of a tree. In autumn, bacteria in the most recently invaded bark become dormant but resume their spread in the following growing season. If the tree is still alive after a year or two, the disease may peter out and the tree recover.

Control: *Therapy:* The disease cannot be cured but, if caught early, the spread of existing infections can be arrested by pruning out infected branches. These must be cut well below the last sign of any staining in the cambial region and when bark surfaces are dry. Between each cut, sterilize the cutting blade by rinsing in or swabbing with a solution of 7 parts methylated spirits to 3 parts water.
Prevention: Infected material should be promptly burnt or buried. Despite

what is said under Significance above, the disease occasionally kills quite large numbers of trees locally. It is therefore advisable to avoid including too high a proportion of very susceptible trees in new planting schemes. Fireblight on outplanted trees is no longer a notifiable disease except where it occurs in or close to premises licensed to trade in susceptible species: see Section 4, p.290.

Remarks: The disease, which originated in the USA, was first described there in 1794; its infective nature was proved in 1871; it was, in 1885, the first plant disease to be shown to be caused by a bacterium. The first record from the British Isles was in 1956/7, from Kent.

Further Reading: Strouts and Patch, 1994.

FROST DAMAGE – SPRING or LATE FROST

Cause: *Freezing temperatures or rapid thawing during active growth*

Damage Type: Killing of actively growing or dehardened tissues.

Symptoms & Diagnosis: *Type (i)*: In spring, many new shoots, leaves or flowers suddenly wilt and within a day or two turn brown or black and shrivel (Fig.140). (Leaves of Buxus sempervirens first turn white – Fig.141). *Type (ii)*: The interveinal tissue of live, fully expanded leaves is torn and irregularly perforated (Fig.142). *Type (iii)*: In summer, small branches or small young trees die.
Confirmation of (i): the damage is confined below a certain height on all affected plants (Fig.143). In valleys and hollows, this height (the frost line) becomes lower as one ascends the valley side – in keeping with the natural accumulation of the coldest air on the

Fig. 140

lowest-lying land. Type (ii) is unlikely to be anything but frost damage. Type (iii): This damage is due to girdling bark death at branch bases or low on thin-barked stems and often coincides

Fig. 141

Fig. 142

Fig. 143

with the presence of narrow dark-brown 'frost' rings in the contiguous portions (Fig. 144). A cross-section cut through dead bark a little below this point showed the time of death to have coincided with the formation of the brown ring). Frost rings may also be present in stems or branches with unin-jured bark.

Additional indicators: Type (i): Frost immediately preceded the onset of symptoms; frost-tender herbaceous and other plants in the vicinity are dam-aged; damage is widespread and on larger trees is confined to the lower parts; later growth is healthy (Fig.145). Type (iii): Particularly severe spring frosts occurred.

Caution: Type (i): On Flowering cher-ry and crabs, cf Monilinia (p.85). Some herbicides also cause similar damage (see p.93). Type (iii): It is necessary to date the injury and rule out bark disease.

Status: Type (i) is common, particular-ly in low lying and flat open areas. Type (ii) and (iii) are probably commoner than the infrequent reports suggest. All types are widespread.

Significance: Type (i): Small, newly planted trees may die; where damage is frequent, survivors are likely to develop stunted, bushy lower branches. Flowering and fruiting can be severely reduced. On large trees, dead shoots

Fig. 144

SECTION **3**

Fig. 145

Fig. 146

disintegrate by summer and damage is concealed by new growth. Type (ii) is a curiosity only. Type (iii): Some species may die if girdled low on the stem.

Susceptible Trees: Type (i): Many are damaged from time to time. Shoots are commonly killed on the following:

Castanea

Fraxinus

Fagus

Juglans

Larix

Picea

Pinus nigra var. maritima

Quercus

Flowers are commonly killed on: Magnolia, Paulownia.

Type (ii): Aesculus, Acer and occasionally other broadleaves.

Resistant Trees: The following are particularly resistant to spring frost:

Betula

Carpinus

Corylus

Pinus sylvestris

Populus

Tilia

Ulmus

Reason for Damage: The freezing of water in dehardened or growing tissues kills cells mainly by dehydration or rupture. The degree of damage depends, among other things, on the inherent resistance of the species, the tree's state of development, on temperature, and on the rate of thawing.

The type (ii) leaf damage originates in the swelling bud: as the undamaged parts expand holes are left where tissue was killed.

Type (iii): In some species (e.g. Thuja, Tsuga) cambial dehardening and growth in spring begins at branch bases and low on the stem well before shoot growth starts.

Control: On known frosty sites, avoid susceptible species. Cover valuable plants when frost threatens until tall enough to escape most damage.

Remarks: The otherwise inexplicable failure of buds to flush, bark cracking, death of small branches, marginal scorch of new leaves or needle tip browning may be due to spring frost.

Spring frost can also cause a remarkable, though rarely reported, curvature of Picea shoots (Fig.146), resembling sub-lethal hormone herbicide damage.

FROST DAMAGE – WINTER COLD and AUTUMN or EARLY FROST	Causes: *(i) Freezing of unhardened tissues (ii) Prolonged low winter temperatures. (iii) Rapid fluctuations from below to above freezing. (iv) Cold soil with drying winds.*

Damage Type: Killing of tissues; bark and wood cracking.

Symptoms & Diagnosis: *Evergreens:* In late winter or early spring and persisting into summer, whole needles or leaves (Fig. 147, Quercus ilex), or their tips or margins, are brown, particularly those produced in the previous year; or foliage is browned on one side of the tree. *All species:* Shoots, twigs, branches or whole stems have died – in deciduous species noticeable only at or after flushing; in summer, bark cankers (Fig.148, Nothofagus obliqua) or longitudinal cracks in the bark and wood or corresponding double callus ridges or 'seams' have appeared.

Fig. 147

Additional indicators: Good recovery shoots develop from below the dead parts (Fig.147); the damage is widespread; the damage has appeared following the type of winter weather described under Causes above.

Caution: Leaf browning and bud and twig death can be confused with de-icing or sea salt spray damage (see p.94). Bark killing can resemble bark disease. Longitudinal bark cracks can be caused by drought (p. 112) or Bacterial wetwood (p. 74).

Status: As common as the weather conditions which induce it.

Significance: Frequently disfiguring, but only the tenderest species in the most extreme winters die or die back severely.

Fig. 148

Susceptible Trees: Mainly exotics from areas with warm winters or with little summer-to-winter temperature variation, e.g. Cordyline australis, Cupressus macrocarpa (Fig.149, contrasting with undamaged × Cupressus leylandii), Eucalyptus, Nothofagus (Fig.148), Pinus radiata, Sequoia sempervirens. But Pinus sylvestris is commonly browned and Populus, Juglans and our native oaks are liable to stem cracking. Late-autumn planted ever-

Fig. 149

Fig. 150

greens are liable to 'physiological drought' in the following winter owing to a paucity of roots. Damage to normally hardy species occurs when they become dehardened during unseasonably mild winter weather.

Resistant Trees: Most of the temperate species, though some native evergreens and some from cold climates,

may still suffer from 'physiological drought' browning.

Reason for Damage: Damagingly low temperatures may occur before frost hardening of tissues is complete in early winter or after partial dehardening has occurred during mild periods in winter. When water in live unhardened or dehardened tissues freezes, cells are killed mainly by dehydration or rupture. Drying winds produce 'physiological drought' damage when water lost from leaves or buds in drying winds cannot be replaced fast enough from cold or frozen ground or through frozen stems to prevent the desiccation of soft tissues. Frost cracks are probably caused by uneven shrinkage of timber on freezing.

Control: *Therapy:* Damage cannot be reversed. The full extent of dieback will not be clear until late in the growing season, after recovery shoots have developed, so any remedial pruning should be delayed until then.
Prevention: Pick species suited to the climate of the region. Protection is not possible for large trees but while they are still small, tender species, especially when newly planted, will benefit from cover or side shelter during adverse conditions. The danger of foliage browning on evergreens can be minimized by spring planting as this allows maximum root development before winter.

Remarks: Some conifers, such as Thuja species, some Juniperus species and Cryptomeria japonica 'Elegans' (Fig.150), assume a bronze, purplish or brownish colour in cold winters but revert to normal in spring. Similar reversible bronzing of yew foliage has been noted after very low temperatures in October. Shoots, unhardened because of a late growing season or leaf cast disease (see p.161) or very close to high pressure sodium street lamps (see p. 7) may be killed in autumn or late summer by early frosts.

| FRUIT TREE TORTRIX on Taxus | Insect: *Ditula angustiorana* (= *Batodes angustiorana*) |

Damage Type: Girdling rings of bark removed from shoots.

Symptoms & Diagnosis: In May the apical 25-40mm of the new shoots turns yellow or golden (Fig.151) and soon after becomes brown and withered, as if caught by frost. There is a sharp division between the green and the discoloured needles.
Confirmation: The shoot is girdled, i.e. ring-barked, where this colour change takes place and there is insect silk, frass and possibly a larva, present.
Caution: Girdling of current shoots *without* silk present can be caused by the vine weevil (Otiorhynchus sulcatus); this species also notches the leaves.

Status: The insect is widespread and common, especially in southern England.

Significance: Damaging on yew hedges and fastigiate cultivars, but only of local and sporadic appearance.

Host Trees: This insect feeds on a wide range of host plants. On trees, killing of the new shoots is recorded on Taxus and once on Picea breweriana; less obvious damage (leaf feeding) also occurs on Pinus, Fagus, Malus, Pyrus, Prunus, Quercus and probably other genera.

Life Cycle: The pale yellow eggs are laid in masses on the leaves in June and July. The larvae hatch in August and feed on the foliage (and on the fruits of rosaceous trees) and, on Taxus, also on the bark of the shoots. They overwinter

Fig. 151

SECTION 3

in silk shelters attached to buds or twigs and recommence feeding in the spring. Pupation is in June, usually on the tree but sometimes on the ground among leaves or detritus spun together with silk. The adult moths emerge in June and July when the males are active and fly around the trees when the sun shines.

Control: Not considered necessary. If damage is conspicuous then handpicking affected shoots, together with the very active larvae, may help reduce pest numbers.

GALLS   Cause: *Mostly insects and mites*

Damage Type: Abnormal growths (Figs 152-164 and others referred to in the text).

Definition: A gall is an abnormal growth reaction of the plant in response

Fig. 152

Fig. 153

Fig. 154

to an intrusive organism such as a bacterium, fungus, nematode (eelworm), mite, insect or possibly a virus. Galls form when plant tissue cells either become abnormally large or are greatly increased in number due to complex biochemical processes (e.g. by the presence or imbalance of auxins or other plant growth regulators). The gall-causing organism is parasitic; it provides the stimulus for the gall to form but is not itself the gall maker. The causal agent may live within or outside the gall tissue. Most galls are of a characteristic form that is specific to the causal organism. Many are of a constant size, although others may exhibit considerable variation. The majority are of annual appearance on the softer leaf tissue; those galls occurring on woody stems, roots and even buds may persist for several years.

Symptoms & Diagnosis: They embody a wide variety of forms: many are protruberances of a characteristic shape, within which the gall causer feeds and reproduces; others are composed of dense patches of hair, usually on the underside of the leaf, and are known as erineum galls; or the gall may take the form of the rolled, and perhaps swollen, edge of the leaf. Some of the most spectacular and large galls occur where a proliferation of axillary buds lead to the formation of a 'witch's broom' growth, many of which are caused by fungi or other microorganisms. Leaf galls and witches' brooms caused by the fungus Taphrina are dealt with on pp.197 and 270 respectively.

Significance: The majority of galls present no risk to the health of the tree. The presence of some species may affect adversely the appearance of the foliage but are otherwise of no concern. Indeed, one or two species, such as the erineum gall on beech leaves that turns red or scarlet in summer, can be considered rather attractive.

Host Trees: Most native genera of broadleaved and coniferous trees are affected by galls, especially on the leaves. Exceptionally there are a few tree genera exotic to Britain that are galled, including Eucalyptus, Gleditsia, Laurus and Picea. In some cases it is only the 'native' species within a genus that are galled. For instance, the white oaks such as Quercus robur and petraea are often heavily galled by cynipid gall wasps (spangle galls, cherry and marble galls, oak apples, etc.) but these are rarely, if ever, found in Britain on North American red oaks such as Q. coccinea and Q. borealis, trees which have their own complex of cynipids in the New World.

Control: Rarely necessary and not usually feasible on large, mature trees. In a few special cases damage is of more concern and in some circumstances control may be considered. See individual entries for details (e.g. Dasineura gleditchiae on Gleditsia p.149; Peach leaf curl, p.197).

Some examples of galls found on trees

Given below are diagnostic descriptions of gall types commonly found on different parts of amenity trees, together with a list of the genera on which they are most often seen.

1. **Galls on foliage**
   a) **Nail and bead galls** – elongated or rounded protruberances on the upper surface of the leaf. Colour sometimes different to leaf lamina. Galls may be smooth or hairy. Most often induced by eriophyid mites, gall midges (Cecidomyiidae) or aphids. Not known to cause harm to the host tree although sometimes sufficiently abundant to be disfigur-

ing. Nail galls occur commonly on Acer, Alnus, Fagus (Fig.152, Hartigiola annulipes) and Tilia (Fig. 153, Eriophyes tiliae). The more globular bead galls are most often found on Acer (Fig.154, Artacris macrorhynchus), Populus (Fig.155, Harmandia globuli) and Ulmus. Bean-like swellings on the foliage of Salix are galls caused by sawflies.

Fig. 155

Fig. 156

b) **Blister galls** – large greenish blisters up to 15mm or more in diameter, often tinged with red, occur on the upper surface of Juglans leaflets. The underside of these galls is lined with white hairs among which live microscopic eriophyid mites (Eriophyes erineus, Fig.156).

Much smaller mite-induced blister galls, often only causing a roughening of the leaf, occur particularly on Pyrus and Sorbus (see p.200). Similar species also affect Acer and Ulmus. Large blister galls on Prunus are caused by the fungus Taphrina deformans (see p.197). See also Rust of box leaves (p.243).

c) **Leaf margin rolled or swollen** – edge of leaf tightly rolled, forming a more or less continuous piping, but sometimes forming a more irregular and intermittent swollen and folded edge. Galls of these types are caused by aphids or gall midges (Acer, Malus, Pyrus, Prunus, Quercus, Salix, Tilia and Ulmus), eriophyid mites (Crataegus, Fagus, Salix and Tilia) and psyllids or suckers (Fraxinus, Laurus).

Fig. 157

d) **Erineum or felt galls** – felt-like or velvety patches of short, dense hairs. Induced by eriophyid mites on either the upper or lower surface. Usually similar colour to leaf lamina at first, but darkening and often becoming brown as the season progresses. One especially striking form of the Eriophyes nervisequus gall, occurring often on purple cultivars of Fagus, is red when first noticed in summer (Fig.157) but later turns brown.

Erineum galls are known to occur on Acer, Aesculus, Alnus, Betula, Crataegus, Fagus, Fraxinus, Malus, Populus (very similar to that caused by the fungus Taphrina populina, see p.197), Sorbus aucuparia and Tilia.

Fig. 158

e) **Cynipid wasp galls** – galls of many different shapes and sizes on Quercus are caused by these insects (also see Remarks below). Probably most familiar and common are the Common spangle gall, Neuroterus quercusbaccarum (Fig.158), found on the underside of leaves in late summer, and the larger cherry gall,

Fig. 159

Fig. 160

Fig. 161

Cynips quercusfolii (Fig.159), on the upper surface in late spring and early summer.

f) **Petiole** – swollen and distorted petioles result mainly from the activity of aphids. Typical on Populus are the purse galls of Pemphigus bursarius (Fig.160) and the spiral galls caused by P.spirothecae, the latter on Populus nigra 'Italica'. The galls cause little or no harm to poplars but one needs to be aware of the former aphid's alternate form that migrates to the roots of various Compositae including thistle (Carduus), sow thistle (Sonchus) and lettuce. Serious damage can be caused to commercial crops of lettuce. However, this is readily overcome by the use of insecticides and modern strains of lettuce resistant to this aphid.

2. **Bud galls**
(1) Enlarged buds that do not flush due to the activity of eriophyid mites occur on Betula and Corylus. These are known as 'big bud' galls (Fig.161 – Corylus damaged by Phytoptus avellanae).
(2) Quercus buds may be galled by a variety of cynipids that cause the buds to change shape and fail to flush. These include Andricus fecundator which causes the artichoke gall; this has many overlapping bud scales around a central core containing the gall wasp larva. Very occasionally heavy infestations in summer may destroy most axillary buds on a tree, resulting in an almost complete lack of lammas growth later that season. Galls of similar appearance occur on Buxus (p.86) and Taxus (p.279) but these are not caused by cynipids.
(3) See also Bud proliferation of Red horse chestnut (p.89).

SECTION **3**

3. Shoot galls

On Picea abies these are partly swollen shoots, rather pineapple-like in appearance when the needles are shortened, as they often are (Fig. 162). These galls are 20mm or more in diameter on vigorous shoots but are often smaller on the lower parts of the crown. They are caused by Adelges abietis and appear from late May onwards. By August or early September the pineapple galls are mature and openings develop through which the adelgids escape. This species, unlike many other adelgids, does not have an alternate generation on a different genus of host tree. Pineapple galls are not a serious problem on amenity trees although they are disfiguring if abundant. When necessary the galls can be cut or picked from ornamental trees as soon as they appear, otherwise no control is required.

Similar galls are caused by Adelges viridis (see p.155); more elongated galls on North American Picea are caused by Adelges cooleyi which has an alternate generation producing white waxy wool on Pseudotsuga menziesii needles (see p.273).

4. Inflorescence

On Fraxinus, swelling and fusion of the pedicels and peduncles forms brownish lumpy masses. These galls, which are sometimes common, are caused by the mite Eriophyes fraxinivorus. They first appear in the spring and remain conspicuous on the tree throughout the year. Currant galls caused by the cynipid Neuroterus quercusbaccarum are common on Quercus flowers (Fig.163). Mossy catkin galls occur on Salix; these are witches' brooms caused by the transformation of the flowers into bunches of vegetative shoots (see p.184).

Fig. 162

Fig. 163

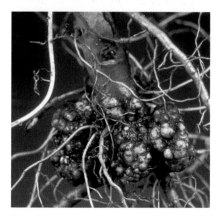

Fig. 164

5. Fruits and Seeds

The only common example is the Knopper gall on the acorn cup of Quercus robur which is widespread in England and Wales and spreading in Scotland (p.153). See also pocket plums on Prunus (p.213).

6. Stem galls

Swellings on twigs occur on Populus and Salix due to the activity of a longhorn beetle or a weevil (see p.255).

Crown gall, typified by one or a number of roughened warty, more or less spherical galls on the stem at ground level (Fig. 164), is caused by the bacterium, Agrobacterium tumefaciens. The bacteria persist in the soil and enter the plant through wounds as diverse as abrasions from tools to damage from eelworm feeding. The tiny young galls are soft and whitish but become hard with age, assume the colour of the tissue from which they grow, and may, depending on their age and the plant's size, reach the size of a football or larger. Within they are a jumbled mixture of woody and soft tissue. Galls may also develop on roots and (often in rows) higher up the stem. If the galls encircle the stem a young tree may be stunted and its flowering and fruiting impaired, but often no harmful effect is apparent. Control measures are therefore unnecessary. Very many species of plant are susceptible but among trees the disease is best known on Malus, Prunus and Pyrus fruit trees. Galls of this nature on Chamaecyparis, Salix and other amenity trees probably have the same cause.

See also Bacterial canker of ash (p.69); Bacterial canker of poplar (p.72); Cherry bark tortrix (p.98); Coatings, deposits, exudations and outgrowths (p.100); Rust of juniper stems (p.244).

Remarks: Galls on Quercus caused by cynipid wasps occur on the leaves and buds, on and below the bark surface of both the aerial parts of the tree and the roots, on the inflorescence and even on the acorn cup. All species have alternating sexual and asexual generations, the first consisting of both male and female cynipids, the other of parthenogenetic females only. The galls caused by these two generations are often very different from each other (see Neuroterus quercusbaccarum Figs.158 and 163): they are usually on different parts of the tree and, in some cases, different species of Quercus (e.g. Knopper gall, p.153). In many species both generations occur in the same year, but a few cynipids take two or more years to complete their life cycle.

Further Reading: Darlington (1968), Meyer (1987), Stubbs (1986).

SECTION 3

GREAT SPRUCE BARK BEETLE

Insect: *Dendroctonus micans*

Damage Type: Destruction of inner bark and cambium on main stems.

Symptoms & Diagnosis: All, or only the upper part of a tree's crown deteriorates and dies; resin bleeds from the main stem (Fig.165).
Confirmation: (i) Protruding from the bark of bleeding stems can be found tubes composed of resin (Fig.166). These may occur anywhere from the roots to the upper crown. They may be single or occur in clusters, leading to an accumulation of resin as large as a clenched fist. Fresh resin tubes indicating recent burrowing activity by beetles vary from white to pale pink or brown and are formed from soft, malleable resin. Tubes containing bark particles and coloured purple-brown to dark brown indicate that the beetle has entered the cambium. Granular resin particles at the tree base indicate an attack below ground level. (ii) Remove the bark around a resin tube. White larvae with brown heads (up to 7mm long) occur in clusters under the bark (Fig.167). The larvae are in brood chambers which can be 30–50cms long and some 10–15cms wide. Within these the larvae pack frass (excreta and other debris) which is interspersed with tunnels and has a 'quilted' appearance (Fig.168). Adult beetles (4–9mm) which vary from pale brown, when immature, to pitchy black may also be found in these chambers (Fig.169).

Status: A serious pest, now established throughout most of eastern, northern and central Wales, the west Midland counties from Shropshire and southwest Staffordshire to the Forest of Dean and parts of Gloucestershire; also in the Trough of Bowland in Lancashire.

Fig. 165

Fig. 166

Fig. 167

Fig. 168

Significance: An often fatal pest which poses a considerable threat to forest plantations but can also occur in amenity plantings such as shelterbelts or single garden trees. Breeds almost exclusively in live and apparently healthy spruce trees.

Host Trees: Most Picea species are susceptible. Picea sitchensis is more severely affected than P. abies. Also, uncommonly, on Pinus sylvestris.

Life Cycle: Female D. micans bore into live bark and excavate a brood chamber where 80–100 eggs are laid. When these hatch, the young larvae feed on living bark at the edge of the brood chamber. The larvae aggregate and feed closely packed side by side. Individual larvae occasionally leave the feeding front to defaecate and they push these resinous faecal pellets back to form concentrations of tightly packed frass which are interspersed with tunnels, giving the quilted appearance mentioned above. Larvae pupate in small, closely spaced, excavations in this frass. Adult beetles cut emergence holes in the thin, papery bark well before they emerge. Some beetles may remain under the bark and form new broods in adjacent areas. Breeding occurs any time between March and October. The length of the life cycle is variable taking from ten months to perhaps two years.

Control: The beetle Rhizophagus grandis, a specific predator of D. micans larvae, has been introduced from Europe and successfully released throughout the infested areas. In addition, many broods are destroyed by woodpeckers. No chemical control is recommended. Timber from infested trees is subject to movement restrictions (see p.290).

Remarks: Introduced accidentally, probably from the Continent, and first found in 1982 in Shropshire. Subsequent surveys showed the insect had been present since 1972.

Further Reading: King & Fielding, (1989), Fielding (1992).

SECTION 3

Fig. 169

GREEN SPRUCE APHID   Insect: *Elatobium abietinum*

Damage Type: Needle browning and defoliation.

Symptoms & Diagnosis: In spring, yellow mottling then browning of the needles is followed, in severe cases, by loss of all but the current needles (Fig 172, Fig.171). Honeydew is plentiful and sooty moulds are present (see p.144). Very occasionally the new shoots show browning and wilting that is similar to frost damage, otherwise current growth is not affected.
Confirmation: There are small green aphids with red eyes (1.5mm long) on the undersides of the needles (Fig.170; use ×10 lens). These are most numerous in spring and early summer, later their moulted skins are often visible on the bark surface of defoliated branches.

Status: Widespread and very common in some years.

Significance: Thin crowns following an attack by E. abietinum give affected trees an unthrifty appearance. A serious economic pest of Christmas tree plantations, which can render trees temporarily unsaleable.

Host Trees: In gardens large Picea abies grown on from Christmas trees are often badly affected, especially in south and east England where they are stressed due to low rainfall. Most North American species are severely affected, notably Picea sitchensis, P. pungens (Blue spruce), and its forms such as glauca × 'Koster', and P. glauca 'Albertiana conica' (dwarf white spruce), though P. breweriana is rarely affected. Some Asiatic species are very resistant but P. asperata is susceptible.

Life Cycle: In Britain, this species breeds essentially by parthenogenesis: the young are borne alive and eggs and male forms are extremely rare. Aphid

Fig. 170

colonies develop in the autumn and breeding continues throughout the winter to peak in May unless checked by severe frost. Winged females are produced in May and June in southern England and Wales, or from June and later into mid summer further north, especially in the Scottish uplands.

Control: On small ornamental spruce use a contact insecticide and spray to run-off (i.e. high volume) as soon as damage is noticed. A routine prophylactic treatment in late August or September using pirimicarb is recommended in those Christmas tree plantations where there is a high risk of damage occurring, for example in the milder parts of the country. [Take note of current Pesticides Regulations – see p.288]

Remarks: E. abietinum is an indigenous European species.

Further Reading: Carter (1972), Carter & Winter (1998).

Fig. 171

Fig. 172

GUIGNARDIA LEAF BLOTCH	Fungus: *Guignardia aesculi* Anamorph: *Phyllosticta sphaeropsidea* *(= P. paviae)*

Damage Type: Killing of leaves and slow defoliation.

Symptoms & Diagnosis: (Fig.173) From July onwards, large chestnut-red or dull brown, irregular blotches are seen on the leaves, often concentrated at the tips and margins of the leaflets, producing a so-called 'marginal scorch' (Fig.174). The blotches are often outlined by a conspicuous yellow band and may here and there be confined by the lateral veins (Fig.174); severely browned leaflets are rolled upwards longitudinally and whole leaves fall prematurely (Fig.175).
Confirmation: Tiny black pimples (pycnidia) develop on the browned parts, mostly on the upper leaf surface (use hand lens).
Caution: Occasionally a Horse chest-nut tree is seen with many of its leaflets browned around the edges but with no yellow-margined blotches. This condition has not been studied closely nor explained but it seems not to be due to Guignardia.

Status: Guignardia leaf blotch is widespread and common in the southern half of England, becoming infrequent northwards.

Significance: Although regarded by many as unsightly, the damage occurs after most growth has taken place so has little effect on the general health of the tree. It is often mistaken for the early onset of autumn coloration but autumn colour changes are fairly uniform over the leaflets, not irregularly blotchy nor markedly marginal.

Fig. 173

Fig. 174

Fig. 175

Fig. 176

Host Trees: Common on Aesculus hippocastanum and A. × carnea. A little damage has been observed on A. indica but without the yellow margins to the brown blotches (Fig.176). Severe blotching has also been recorded in the USA on A. hippocastanum 'Baumanii' and A. × carnea 'Briotii' so can be expected to occur in this country too.

Resistant Trees: In the USA, A. turbinata is said to be rarely affected. No information is available on other species grown in this country.

Infection & Development: It is supposed that the fungus overwinters on fallen leaves and that in spring the Guignardia state develops on these to produce ascospores which infect the new leaves, as happens in the USA. This remains to be confirmed, however, as the Guignardia state has rarely been recorded in Britain. In summer, conidia produced on the leaf blotches cause further leaf infections.

Control: is not necessary on outplanted trees.

Remarks: The disease probably originates from North America. It has been known in this country since at least 1935.

HOLLY LEAF MINER   Insect: *Phytomyza ilicis*

Damage Type: Removal of internal leaf tissue.

Symptoms & Diagnosis: On the leaves there is an irregular creamy-green to yellowish blotch with a brown or purplish-brown central area (Fig. 177). This whole area is raised slightly above the leaf surface, most often on the topside of the leaf but also on the underside on some thick-leaved cultivars. *Confirmation:* If the raised epidermis is pierced or removed, a cavity (a mine) is revealed in the thickness of the leaf.

Fig. 177

Status: Widespread, sometimes very common.

Significance: Disfiguring on ornamentals, particularly as the mines are most common on the lower and more visible leaves, i.e. from ground level up to about eye level, than they are higher up on the tree. Damaged leaves persist on the tree throughout the year.

Host Trees: Ilex species.

Life Cycle: Adult flies occur in May and June when they lay eggs on the mid-rib of the leaf. The larvae hatch quickly and feed by mining the mid-rib before moving into the leaf lamina. These mines do not become visible as leaf blotches until the winter. By January they are about 4–6mm in diameter and similar in appearance to, but darker in colour than, mature mines. Further expansion of the mines continues until March or April when each larva pupates just below the epidermis, with the tip of the puparium partially visible through a hole in the leaf surface. The adult emerges through this round hole in early summer.

Control: No chemicals are recommended. Handpicking and destroying infested leaves in the spring (March/April) will reduce the population.

HONEYDEW AND SOOTY MOULDS — Cause: *Sap-sucking insects*

Damage Type and Significance: Honeydew and Sooty moulds do not themselves harm the tree but are the result of feeding by sucking insects which may be harmful (see also Remarks below).

Symptoms & Diagnosis: *(i) Honeydew:* Sticky deposits on the tree; also on the ground, plants and other objects below the canopy. Honeydew is phloem sap, minus some amino acids and sugars, excreted by sap-sucking insects including aphids, adelgids, phylloxerids, coccids and soft scales (but not by leafhoppers and rarely by armoured scales). Ants and wasps are often indicative of honeydew deposits. Wasps are regular late summer visitors to cypresses infested by Cinara cupressi and to Salix where the giant willow aphid (Tuberolachnus salignus) is present on the branches. *(ii) Sooty moulds:* Honeydew often becomes colonised by saprophytic black coloured moulds as the season progresses, more particularly in dry years when rain has not washed the deposits away (Fig.178). These 'Sooty moulds' can greatly reduce the amount of light reaching the leaf so that the normal photosynthetic process is impaired. Softer leaves may become desiccated (appear scorched) and will fall prematurely during long periods of dry weather, apparently not a direct effect of the insects' feeding.

Host Trees: Various broadleaves and conifers. Problems with honeydew falling on pavements, motor vehicles and gardens are most frequently report-

Fig. 178

ed for Tilia (species without hairs or glands on the leaves), Acer pseudoplatanus, Fagus sylvatica (p.80), Betula, Quercus, Prunus, Salix, Abies (p.251), Picea (p.140), Pinus, Cedrus (p.254), Cupressocyparis (p.109) and Juniperus and may be experienced with Pittosporum (p.48).

Control: As the production of Honeydew and Sooty moulds depends on the presence of certain insects (refer to the relevant entry given under Host Trees above), prevention depends on the control of the insect. Honeydew can be removed from hard surfaces by washing or scrubbing with ample, preferably warm, soapy water.

Remarks: Honeydew is essentially just sugar-water; it can be unpleasant and a nuisance but it is harmless to people, textiles and vehicle paintwork. Occasion-

ally a thick honeydew deposit followed by a little rain can become slippery underfoot.

Further Reading: Bevan (1987); Carter (1992).

HONEY FUNGUS   Fungi: *Armillaria species*

Damage Type: Root and root collar killing; root and butt rotting.

Symptoms & Diagnosis: A tree (i) has suddenly died or has died after a short or lengthy period of increasing ill health (Fig.179), or (ii) shows a general deterioration in crown condition indicative of a root or root collar problem (Fig.180); as described in Table III, p.14, or (iii) has blown over to reveal rotted roots.
Additional indicators: Other trees or shrubs in the immediate vicinity have died over a number of years. The recent

Fig. 179

death is another in a series of deaths of adjacent trees in a hedge or close-planted row (Fig.181). Resin, gum or a watery liquid exudes from the lower stem of the affected tree (Fig.182). Clumps of Armillaria toadstools (see below) appear at or near the tree base or on nearby stumps (Fig.183).
Confirmation: Bark at the root collar and/or of roots is dead and when cut away reveals thin, white or creamy white, paper-thin, cohesive sheets of fungal tissue (mycelium), sometimes with fan-like striations, sandwiched between it and the underlying wood (Fig.184). The mycelium may also marble the thickness of the dead bark. Infected bark usually has a strong 'mushroomy' or fungoid smell. Dead roots may be decayed: the rot is stringy and distinctly wet.
Honey fungus toadstools (Figs.183, 316) grow mostly in clumps and only in autumn. They are brown or honey-yellow-brown, with roughly the stature and texture of tall cultivated mushrooms. The stalks always have a whitish collar near the cap; the gills are clearly joined to the stalk; the spore powder is white or cream and often conspicuous on surfaces beneath the caps. (If not, place a cap, gills down, on white paper for a few hours. If the spore print is not white or cream, it is not Honey fungus).
Caution: (i) 'Honey fungus' comprises several species, not all of them strongly pathogenic. One in particular, Armillaria gallica (= A. bulbosa), offers

SECTION **3**

little threat to healthy trees but often invades trees dying or severely stressed from some other cause, e.g. drought, soil compaction, Phytophthora root disease. It is also a very common inhabitant of stumps of trees and shrubs and of other buried wood, and its conspicuous, abundant, tough rhizomorphs (Fig.185: black, bootlace-like strands) can often be found ramifying through leaf litter, compost heaps, in the soil and beneath tree bark killed by, for example, fire or lightning. For this reason the presence of rhizomorphs (or Armillaria toadstools) alone should not be taken to indicate that the ill health of trees in the vicinity is due to Honey fungus; their discovery in the absence of clear signs of Honey fungus *infection* does not warrant the institution of control measures.

(ii) **White root rot**, another root- and root collar-killing disease, caused by the fungus Rosellinia necatrix (anamorph Dematophora necatrix) has occasionally been recorded on trees in SW England and the Scilly Isles. Small pockets of white mycelium are scattered through the killed bark and these might be mistaken for Honey fungus, but the extensive sheets of mycelium beneath the bark, so characteristic of Honey fungus, are absent. White root rot rots only fine roots, not woody roots, and produces a cobweb-like white or greyish mycelium over the killed bark and in, sometimes on, the soil. All roots of infected trees should be removed and replanting with susceptible species (Malus, Ligustrum, Morus, Populus, Prunus (cherries) and Pyrus) avoided.

Status: Very common and widespread on wooded or previously wooded sites. The disease is very uncommon in urban roadside trees.

Significance: With the exception of epidemics such as Dutch elm disease, Honey fungus probably kills and debilitates more amenity trees in this country than any other single living agent.

Host Trees: An extremely wide range of woody plants are attacked though susceptibility varies greatly. Among the commoner ornamentals, the following seem to be particularly susceptible: Araucaria, Betula, Cedrus, Chamaecyparis, × Cupressocyparis, Juglans regia, Ligustrum, Malus, Prunus (the Flowering cherries), Salix, Sequoiadendron and Thuja. Many shrubs are susceptible, notably Syringa (lilac) and the two privets commonly used for hedging: Ligustrum ovalifolium and L. vulgare.

  Resistant Trees: Acer negundo, Juglans hindsii, Juglans nigra and Taxus baccata are highly resistant if not immune.

Trees with a useful degree of resistance are:

Abies species

Ailanthus altissima

Buxus sempervirens

Calocedrus decurrens

Carpinus betulus

Catalpa bignoniodes

Crataegus species

Fagus sylvatica

Fraxinus excelsior

Ilex aquifolium

Juniperus species

Larix species

Liquidambar styraciflua

Nothofagus species

Platanus × hispanica

Prunus laurocerasus

Prunus spinosa

Pseudotsuga menziesii

Quercus species

Robinia pseudoacacia

Tilia species

Infection & Development: The fungus resides in buried wood (stumps, posts etc.) and in dead roots on living

Fig. 180. Two lime trees of the same age and size at planting, the retarded growth of one due to Honey fungus infection.

Fig. 181

Fig. 182

Fig. 183

SECTION 3

Fig. 184

Fig. 185

trees. It spreads locally by growing through the soil and leaf litter as strands (rhizomorphs, Fig.185) whose tips can infect undamaged, live bark of roots and the root collar, or dead wood. The fungus grows through and kills the phloem and cambium and later invades and decays the wood. The tree dies once it is girdled at the root collar or as the result of extensive root killing. Some trees blow over while still alive, as the result of extensive root decay. On rare occasions, airborne spores also give rise to new infections.

Control: *Therapy:* None available.
Prevention: To prevent continuing spread of the disease, ideally (a) dig out all infected material, then replant. In hedges and screens, also remove the apparently healthy plant next to the one showing symptoms, in both directions. Otherwise (b) dig out as much infected material as practicable and replant with resistant species or (c) obstruct rhizomorph spread by inserting durable plastic sheeting barriers in the ground between the source of infection and the plants to be protected (see Further

Reading for details). If deaths continue, (d) check diagnosis and, if appropriate, repeat (a) or (b). Where none of the above control measures are practicable, (e) plant with immune species, or accept the likely further occasional losses, removing plants as they die; or relocate the tree-growing area and use the infected area for another purpose (herbaceous plants, for example).

The toadstools are of no practical significance in the spread of the disease.

The value of available chemical control measures is dubious although recent research has shown certain soil fumigants to be useful in killing the fungus in tree stumps.

Remarks: Plants 30 metres or more from infected trees or stumps are not at risk.

The pathogenicity of Honey fungus was first demonstrated by the illustrious German 'Father of Forest Pathology', Robert Hartig, in 1873.

Further Reading: Greig, Gregory and Strouts, 1991.

HONEYLOCUST GALL MIDGE

Insect: *Dasineura gleditchiae*

Damage Type: Distortion of leaves and shoots.

Symptoms & Diagnosis: Leaflets are transformed into yellowish-green or reddish pod-like structures (Fig.186); the shoots may also be distorted.

Status: Occurs in south-east England but its distribution is incompletely known. It is likely to spread to wherever the host tree is grown.

Significance: Can ruin the appearance of foliage. New growth may cease by the end of July.

Host Trees: Gleditsia triacanthos 'Sunburst'. G. triacanthos 'Elegantissima' is reported to be resistant.

Life Cycle: Adult midges emerge from the soil in May or later. The eggs are laid on leaflets. These develop into pod-like galls, each containing a number of gregarious orange larvae up to 3.5mm long. Development takes only 3–4 weeks from egg to adult. There will be several generations during the summer. The larvae pupate in the galls. However, larvae from the last autumn generation return to the soil to overwinter before pupating in spring. Some may remain in the soil for more than one year.

Fig. 186

Control: A difficult species to eradicate due to its protected habitat in the galls and to the diapausing larvae in the soil. A suitable contact insecticide applied to the soil in spring before the first adults emerge may help reduce populations. [Take note of current Pesticides Regulations – see p.288]

Remarks: D. gleditchiae originated in N. America. It was first reported in Britain in the early 1980s, possibly having been introduced from Europe, where it was first found in Holland. It is now established in several countries.

SECTION **3**

HORSE CHESTNUT SCALE

Insect: *Pulvinaria regalis*

Damage Type: Growth reduction but primarily an aesthetic problem.

Symptoms & Diagnosis: In April and May, scattered or very numerous and crowded circular, white spots, or larger masses of white wax topped with a flattened brown or orangy scale appear on trunks or large branches and persist throughout the summer (Fig.187).
Additional indicator: From August to October, honeydew and Sooty moulds appear on the leaves in small quantities.
Caution: Large quantities of honeydew on Tilia or Acer pseudoplatanus at this time of year more probably indicate the presence of aphids (see p.144).

Status: Southern Britain north to Lancashire and west to Glamorgan. Also found in Perthshire in 1998 and in Dublin. Likely to spread to the whole of Great Britain and Ireland.

Significance: Only a problem on trees in urban situations. Often very abundant when first noticed in a fresh locality; numbers may subsequently decline over several years, possibly due to bird predation of nymphs overwintering on the twigs. High P. regalis populations have been shown recently to cause a significant effect on the roots of Tilia cordata (35% reduction in the root/shoot dry weight ratio). Similar effects were also noted on Aesculus and Acer pseudoplatanus. Reductions in stem diameter and shoot elongation were also noted on these last two hosts. This investigation involved 4 year old potted plants but did suggest that a significant effect will occur on larger trees, especially where these are stressed due to lack of water, oxygen or nutrients.

Fig. 187

Host Trees: Most often reported on Tilia spp. including T. cordata, T. × europea and T. platyphyllos; also Aesculus hippocastanum and Acer pseudoplatanus. Has been found on a wide variety of other broadleaved trees including other Acer spp., Cornus, Laurus, Magnolia, Populus, Prunus, Salix and Ulmus.

Life Cycle: The female scales, which are 5–7mm long, move onto the trunk from the branches in late spring. The rare male scales are smaller (3mm long) and are found only on the branches in April and May. Many females are parthenogenetic (do not require fertilization) and lay up to 3,000 eggs in May from which the tiny yellowish 'crawlers', 0.5mm long, hatch in summer and migrate to the leaves; they are sometimes visible as a moving carpet of dust. These crawlers or nymphs settle and feed on the underside of the leaves during the summer (Fig.188), moving back to the twigs in autumn where they

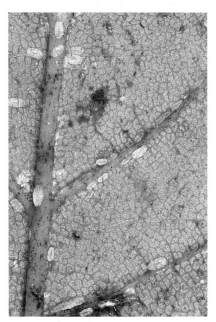

Fig. 188

continue feeding throughout the winter. By March or April they are fully grown and the females then migrate back onto the bole of the tree to lay eggs. It is these brown, dead female scales surrounded by white, waxy wool providing protection for the eggs until they hatch which attract attention.

Control: Not recommended, except on young trees in exceptional circumstances when the use of a contact insecticide may be considered. [Take note of current Pesticides Regulations – see p.288] Application to the underside of the leaves will need to coincide with the presence of the leaf feeding stage in the summer. Scrubbing the bark with a mild detergent solution will improve the appearance of the tree but is of doubtful value in controlling this particular scale insect.

Remarks: P. regalis was first found in Britain in 1964 having arrived from overseas. Unknown in Europe before the 1960s. The origin of this species outside Europe is not known.

Further Reading: Wainhouse (1994)

JUNIPER SHOOT MOTH   Insect: *Argyresthia dilectella*

Damage Type: Hollowing out of scale leaves and shoot tips.

Symptoms & Diagnosis: In spring scattered shoot tips turn purple at first and then become brown. Such damage is scattered throughout the crown.
Confirmation: The affected scale leaves and shoot tips are hollow. In these look for small entrance/exit holes and also small silk cocoons on the underside of the foliage (Fig.189).
Caution: Resembles damage from the fungus Kabatina (Fig.190) but there the shoots are not hollowed out.

Status: Not fully known but may occur wherever junipers are grown.

Significance: Disfiguring on ornamental trees; in extreme cases may severely reduce new growth (noted thus on × Cupressocyparis).

Host Trees: Juniperus; also on × Cupressocyparis and Chamaecyparis.

Life Cycle: Adult moths are small, 8-10mm wingspan, pale silver in colour with yellow-brown markings. They lay their eggs on the shoots in July. These soon hatch and the larvae tunnel in the shoots, feeding throughout the winter to complete their development by May. Full grown larvae leave the mines and pupate in spindle-shaped silk cocoons about 5mm long on the underside of the foliage in June and July.

Control: If attacks are recurrent and severe, spray affected trees with a suitable insecticide such as diflubenzuron when eggs hatch, to prevent the larvae

Fig. 189

boring into the shoots. [Take note of current Pesticides Regulations – see p.288]

Remarks: Two other species occur in Europe, A. trifasciata on Juniperus, Thuja and various other Cupressaceae and A. thuiella on Thuja occidentalis and on cypresses but not on Juniperus. A. trifasciata has recently been found in Britain and A. thujiella will also probably soon reach this country.

KABATINA SHOOT BLIGHT

Fungi: *Kabatina thujae*
Kabatina juniperi

Fig. 190

Fig. 191

Damage Type: Shoot killing.

Symptoms & Diagnosis: Scattered one-year-old shoots are faded, yellowed or browned (Fig.190) .
Confirmation: The basal centimetre or two of affected shoots is constricted, dead, and grey or greyish brown in colour (Fig.191). On this portion, black, 1–2mm diameter pin head-like structures (fruit bodies) may be visible, bursting through the epidermis, or do develop if the shoot is kept wrapped in damp paper for a few days (Fig.192). Non-girdling lesions several centimetres long may be present.
Caution: Resembles damage from the moth Argyresthia (p.151), but there the shoot is hollowed out. Except on Chamaecyparis, Kabatina shoot blight could be mistaken for minor damage from Seiridium (p.107) or Phomopsis (p.201).

Status: Occasional. Probably widespread.

Significance: Outside nurseries, this is a curiosity only. It rarely causes noteworthy damage and even this is soon obscured by new, healthy growth in the same or the following year.

Host Trees: Confined to the Cupressaceae (cypress family). Quite severe damage (numerous dead shoots) has been seen on Cupressus macrocarpa 'Lutea' and × Cupressocyparis leylandii up to 12 ft high.

Infection & Development: The fungus overwinters in dead shoots. In spring it produces spores which infect the developing shoots, apparently only through wounds of some kind. The fungus spreads in the shoot perhaps far enough to girdle it (Fig.191), and then fruits, but no further spread occurs in the second year.

Control: None required. Dead shoots can be clipped off to improve the plant's appearance.

Fig. 192

Remarks: Dead foliage of cypresses and other Cupressaceae is often colonized by species of Pestalotia; this fungus also kills foliage kept moist and heavily shaded. Its fruit bodies can be mistaken for Kabatina but those of Kabatina are typically confined to the girdled part of the shoot while Pestalotia fruits anywhere on the leaves and shoots.

Kabatina and the disease it causes was described first from Germany in 1964 and first recognised in this country in 1970, in the south of England.

KNOPPER GALL   Insect: *Andricus quercuscalicis*

Damage Type: Deformation of acorn cup; abortion of acorn.

Symptoms & Diagnosis: In July and August an irregular, ridged, cone-shaped gall, c. 20–30mm diameter × 15–20mm high, with a hole in top centre, develops attached to the acorn cup (Fig.193). The acorn is absent or poorly developed. These galls are green and have a sticky coating that gives them a glossy appearance. They turn brown in late August or early September when they fall from the tree, generally before the healthy acorns. Several galls may develop on one cup, although they are most often single.
Confirmation: Cut open the gall to reveal the white larva within the cell at its base (Fig.194).

Status: Widespread and common in England, Wales and spreading northwards into Scotland. Also in the Channel Islands.

Significance: Acorns abort although occasionally a deformed seed will develop and may still germinate. Knopper galls can be of local importance in seed stands where, in poor mast years, a high

Fig. 193

percentage of the acorns may be affected. There is no risk to the health of individual trees or to long-term supplies of acorns as, in good mast years, a smaller percentage of acorns are affected by Knopper galls.

Host Trees: Quercus robur. The galls of the alternate generation of A. quercuscalicis are on Q. cerris in spring (Fig.195, and see below).

Life Cycle: The eggs of this cynipid wasp are laid into the developing acorn

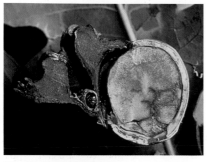

Fig. 194

cups in June. Each larva develops singly in a cell at the base of the gall and overwinters in the fallen gall in the litter. Adults (parthenogenetic females) emerge in the following February, although a proportion of the population remains in the galls for 2 or even 3 years. The females fly to Q. cerris where they lay eggs into the male flower buds (Fig.196), causing tiny galls to form on the catkins (Fig. 195). From these galls both male and female A. quercuscalicis emerge and return to Q. robur in June to initiate the next generation of knopper galls.

Control: Not necessary or recommended. The removal of Q. cerris, as is sometimes suggested, is ineffective since the females fly long distances.

Remarks: The first confirmed record in Britain was in 1961 (Northamptonshire) but the species may have been here earlier.

Further Reading: Jukes (1984).

Fig. 195

Fig. 196

LARCH WOOLLY APHIDS (ADELGIDS)

Insect: *Adelges laricis*

Fig. 197

Fig. 199

Fig. 198

Damage Type: Needle distortion, shoot dieback; death of branches or whole trees.

Symptoms & Diagnosis: The foliage appears bluish white (with a woolly secretion) and the needles of the dwarf and long shoots are bent or distorted (Fig.197). Copious honeydew is present and becomes blackened with Sooty moulds (p.144). Later, needles turn purplish brown and drop, and shoots die back (Fig.198).

Confirmation: Adelgids in the wool are dark purplish-grey.

Caution: Needles of the dwarf shoots with a double elbowed kink and slight yellowish swelling, associated with blobs of white waxy wool and green aphids, have been damaged by A. viridis.

Status: A. laricis is common and widespread; A. viridis is also widespread but occurs far less frequently.

Significance: A. laricis can be a serious pest of European larch. Small colonies can build up into vast numbers in a few years. On young trees attacks will last two or three years resulting in weak long shoot growth and poor bud formation. This is followed by poor flushing or complete failure of buds the following year causing shoot dieback which sometimes weakens and kills the whole tree. This condition is known as Larch dieback.

A. viridis causes no serious damage.

Host Trees: A. laricis occurs as a pest on Larix decidua and the Alpine strains are highly susceptible; Sudeten provenances are very resistant. L. kaempferi

SECTION **3**

and L. × eurolepis are not favoured by this insect. A. viridis occurs on L. decidua and L. × eurolepis.

Life Cycle: A. laricis is present on larch needles from May to September. The nymphs that overwinter are black and remain near buds or in crevices on the current shoots until the spring. During this period they do not secrete any wax. Winged generations give rise to galls during the following year on the current shoots of Picea. The galls are spherical (1.0–1.5 cm diameter), creamy in colour with a waxy texture (Fig.199). Affected shoots often continue to develop beyond the galls.

Control: No reliable treatment can be recommended for large trees. On small trees of high value A. laricis may be sprayed with a permitted contact insecticide. [Take note of current Pesticides Regulations – see p.288] Treatment should be applied on a suitably mild and calm day between November and February.

LEAF AND FLEA BEETLES

Insects: *Various species in the family Chrysomelidae*

Damage Type: Removal of the leaf epidermis and partial leaf killing.

Symptoms & Diagnosis: Brown to black patches appear on leaves.
Confirmation: The damaged areas are seen to be translucent when held up against the light (Fig.200). Close examination reveals that the lower epidermis has been removed, leaving a skeleton of veins with the upper epidermis dead. The damage is most noticeable in late summer. Beetles or larvae may be found feeding on the leaves. The adult beetles are mainly metallic green, bronze or blue, but Chrysomela populi is red and black. The larvae vary from whitish or pale yellow to dirty grey with dark heads and many shiny black plates on the body. They are mostly very messy feeders: wet excreta and moulted skins contaminate the leaf surface, making the foliage unpleasant to handle.

Status: Widespread and common.

Significance: Infestations spoil the appearance of amenity trees but cause no serious long-term damage. Damage by flea beetles (Chalcoides spp) is restricted mainly to nursery stock.

Host Trees: Leaf beetle damage occurs most often on Populus, Salix and Alnus. A number of different species occur on these hosts of which the following are encountered most frequently.

Common on Salix and Populus:
beetles greenish or bronze-coloured – Phyllodecta vitellinae.

Less common on Salix and Populus:
beetles bluish – Phyllodecta vulgatissima.
beetles red and black – Chrysomela populi.

Mainly on Salix:
beetles blue – Plagiodera versicolora.

On Alnus:
beetles coppery blue or metallic green – Chrysomela aenea.

Other species of leaf beetle occur on Betula, Corylus and Crataegus but do not cause serious damage and are rarely seen.

Fig. 200

Life Cycles: All species overwinter as adults, either under loose bark, in trunk crevices or in the litter/soil layer. The beetles become active in the spring, when they feed on newly flushed leaves and bursting buds. Eggs are laid on the underside of the leaves where the larvae feed gregariously at first, but later often singly. The larvae drop to the soil to pupate. There are normally two generations per year, the second being far more abundant.

Control: Only justified in nurseries.

Further Reading: Jobling (1990).

LEAFHOPPERS

Insects: *Order Hemiptera.*
Mostly in the sub-family Typhlocybinae

SECTION 3

Damage Type: Removal of the contents and consequent death of the mesophyll cells of the leaf.

Symptoms & Diagnosis: In early summer, white or yellowish spots are seen on the upper surface of the leaves. In later summer, larger, irregular areas of discolouration are seen, giving the foliage a distinctive mottled, bronzed or silvered appearance (Figs.201,202).
Confirmation: The undersides of the leaves harbour leafhopper adults (these are very active, leap off and fly away), feeding nymphs or moulted skins (before moulting, a nymph inserts its rostrum into the leaf tissue to feed; long after it has moulted and gone, the ghostly slender white skin remains, attached to the lower epidermis).
Caution: Damage that is very similar in appearance can be caused by spider mites. With ×10 lens look for the presence of fine silk and/or tiny pale cast skins (spider-like) on the underside of the leaves (see also the Lime mite, p.172). No silk will be present on leaves damaged by leafhoppers.

Fig. 201

Fig. 202

Status: Widespread and common although the damage is rarely reported.

Significance: Only an aesthetic problem but some species may occasionally be very common on individual trees, when much of the foliage will be affected. The leaves may appear unsightly after an attack but the health of the tree is not seriously affected.

Host Trees: A wide variety of broadleaved trees. The Acers, especially A. pseudoplatanus, are particularly susceptible; Carpinus, Fagus, Quercus Tilia and many rosaceous species are also attacked.

Life Cycle: Most tree-feeding leaf-hoppers have either one or two generations per year and overwinter as eggs laid just below the bark on young shoots. A few species that overwinter as adults have only one generation per year and these normally overwinter in the thick cover provided by evergreens such as holly, ivy and conifers, returning to a broadleaved host to lay their eggs in the spring.

Control: Unnecessary.

Further Reading: Alford (1991).

LEAF MINERS

Various insect larvae; *mainly Lepidoptera, Coleoptera, Hymenoptera and Diptera*

Damage Type: Removal of internal leaf tissue and partial leaf killing

Symptoms & Diagnosis: (Figs.203–206) Discoloured patches occur on the leaves and are usually clearly defined and often of a characteristic form. On broadleaved trees they may affect either upper or lower epidermis, rarely both. Some leaf mines form an obvious blister, others are in the same plane as the leaf. The mines may take the shape of an irregular blotch, sometimes confined between the lateral veins, or have a definite serpentine appearance (linear mines). The position of the mine on the leaf and dark lines of frass within it may be important diagnostic characters. Discrete circular or irregular mines, especially on Quercus, are often caused by case-bearing larvae of small moths (Coleophora spp.). Mines on conifers usually occupy the whole needle width, but may be in various positions along its length. Larix needles mined by the larch casebearer (Coleophora laricella) give the appearance of frost damage. On cypress-type foliage whole shoot tips

are often hollowed out by the larvae (p.151). More extensive mining by some insects can give the foliage the appearance of having been scorched by fire (see Laburnum below).
Confirmation: Mined leaves and needles are mostly translucent when held up to a bright light. The feeding insect may be visible in the leaf. On broadleaves the epidermis can easily be separated and peeled back to reveal the larva and its excreta within the mine.

Significance and Host Trees: Leaf miners occur on most trees in Britain. Only the commonest and most striking are dealt with in this book. Severe infestations on ornamental broadleaves are disfiguring but do not affect the health of the trees significantly. Examples include Fagus (p.79), Ilex (p.143), and those described below. Damage to most conifers with needles occurs on a large scale in plantations only. Species with cypress-like foliage may be disfigured by Argyresthia (p.151).

Control: Not necessary in most situations. However, in special cases leaf miners can be controlled with a systemic insecticide approved for use on amenity trees. [Take note of current Pesticides Regulations – see p.288] Where control is considered desirable, the identification of the insect and a knowledge of its life cycle are essential before an insecticide can be recommended.

LABURNUM:
Agromyzid fly	*Agromyza demeijerei*
Laburnum leaf miner (moth)	*Leucoptera laburnella*

Both species cause extensive blotch mines (Fig.203). Those caused by A. demeijerei turn brown giving the leaf a scorched appearance. L. laburnella mines are more blister-like and contain lines of black frass. Both insects have two generations per year in June/July, and then August (L. laburnella) or September/October (A. demeijerei). Control is not usually recommended but on small trees the use of a systemic insecticide as soon as damage starts may be beneficial. First generation larvae of L. laburnella pupate in small white cocoons on the underside of the leaves, otherwise the larvae of both species fall to the ground and pupate in the litter.

QUERCUS ILEX:
Zeller's midget moth	*Phyllonorycter messaniella*

Small brown blotch mines with 4–5 per leaf are sometimes abundant in the winter. Mines persist and affect the appearance of the foliage throughout the year (Fig.204). Often the mines are torn open by birds searching for the larvae. Damage is of aesthetic importance only, the health of the trees will not be affected and no control measures are necessary. P. messaniella also occurs on deciduous oaks during the summer but these mines are not noticeable or detrimental.

Remarks: More than 40 species of leafminers are known to occur on Quercus in Britain, mostly on Q. robur and Q. petraea, a few on Q. cerris and four on Q. ilex. None of these have a serious affect on the health of the trees.

SECTION 3

Fig. 204

Fig. 203

PLATANUS:
Moth larva   *Phyllonorycter platani*

Small but distinct blotch mines visible from below the tree, with as many as eight per leaf. These mines are caused by the larvae of P. platani, a very small moth that has spread north and westwards through Europe this century and arrived in west London in 1990. The collection and removal to other locations of fallen leaves which contain overwintering pupae in the mines will result in the spread of this species outside its current locations. Not damaging to mature trees and no control recommended.

PICEA and ABIES:
Tortricid moths   *Epinotia spp.*

Spruce needles are mined or grooved by the larvae and become either partially or completely transparent. Often the loose damaged needles are spun together with obvious silk and frass (Fig.206). In particular the larvae of the Spruce bell moth (E. tedella) can affect large numbers of needles on Picea in the autumn (Fig.205). Similar but less untidy damage on Abies grandis and A.procera is caused in June and July by

E.subsequana. With both pests, the damaged needles are evident during the winter and may not fall until the following spring. Rarely a serious problem on amenity trees but damage may occur if these are close to spruce plantations. Other Epinotia species cause more discrete patches in late spring and summer, often involving just a few needles. No control is required for any Epinotia species.

Fig. 206

Fig. 205

LEAF SPOTS AND BLOTCHES (fungal)
(see also Scab diseases, p.245)

Fungi: *Various species*

Damage Type: Leaf killing; some-times defoliation and shoot killing.

Symptoms & Diagnosis: (Figs.207–221) *Leaves* (and perhaps other soft tis-sues) are variously spotted or blotched, or completely or partially yellowed or browned, and may fall prematurely.

The spots or discoloured areas may be separated from healthy green tissue by a band of a third colour; their margin may be sharp or diffuse or appear to be furnished with a fringe; spots may be concentrically zoned.

Infected leaves may become distorted (Fig.220); the dead tissue may disinte-grate to leave small holes (a symptom often described as 'shot-holes') or ragged leaf margins; large portions of spotted leaves may become yellow.

Infected leaves may be retained on the tree for a long time or fall prematurely – in some diseases, even when still largely green; leaves may fall from any-where in the crown, but in some dis-eases defoliation characteristically begins in the lower branches and progresses up the tree as the season wears on.

New shoots and *fruitlets* as well as the leaves may (depending on the disease in question) be spotted. The lesions may extend right round the shoot and cause the part distal to the girdle to die; fruitlets may become deformed and may fall before ripening.

Confirmation: On the discoloured tis-sue on one side of the leaf or the other, perhaps chiefly on the veins, tiny pim-ples or blisters (fruit bodies) may be visible (a hand lens is usually neces-sary); when moist, these may extrude a blob or tendril of a paste-like or mucilagenous substance (spore masses).

Additional indicators: Severe out-breaks follow periods of wet, warm weather. Older twigs may bear small,

healed lesions and dead tips from infec-tions in previous years.

Caution: Similar symptoms may result from bacterial infections, pest infesta-tions, herbicides (p.93), drought (p.112), winter cold (p.129), or hail (p.179). Brown spots may also develop where leaves have been punctured by thrash-ing against spines of their own or of neighbouring plants (use hand lens).

Status: See the notes on specific dis-eases below.

Significance: This depends on the severity and frequency of defoliation and on whether shoots are killed. Outbreaks of many of these diseases can be disregarded as they occur only spo-radically and so are only temporarily disfiguring.

If severe defoliation results in the pro-duction of a new flush of shoots very late in the season, these will not harden completely and are likely to be killed by winter cold.

Severe defoliation in several successive years will reduce the tree's vigour and its capacity to withstand attacks by pests and fungi, resulting in various degrees of dieback and even ultimately in the tree's death.

Repeated attacks of Marssonina on Weeping willow may spoil the tree's weeping habit to such an extent that control becomes desirable.

Host Trees: Most species are liable to one or more such diseases; only the commonest, most striking or most dam-aging are dealt with in this book (see below); a few particularly notable ones have been given a separate entry.

Infection & Development: With exceptions mentioned below, the causal fungi overwinter in the infected fallen

SECTION **3**

leaves. In spring, spores produced from these are carried to newly developing soft tissues in air currents or in wind-blown or splashed rain, depending on the fungus involved, and, if the weather is suitably wet and warm, cause the first (primary) infections of the season. Spores produced from resultant lesions cause further (secondary) infections during the growing season. The severity of the diseases depends partly on the susceptibility of the host tree and partly on the prevalence of weather wet and warm enough to allow abundant spore production, dispersal and germination.

Control: Although all these diseases are theoretically preventable by maintaining a coating of fungicide on the new, susceptible tissues, only in the case of repeated attacks from Marssonina on valuable weeping willows or scab on crab-apples is this ever likely to be warranted.

To reduce the risk of re-infection in the following spring, it is often recommended that the fallen, infected leaves are collected up and removed from the site or destroyed. Unless this can be done very thoroughly, it is probably not worthwhile as, in suitable weather conditions, only very few primary infections are necessary to set off a severe outbreak. Similarly, the logical recommendation to remove infected twigs or fruit when these harbour the fungus over winter is likely to be practicable and therefore effective only on the smallest of trees.

REMARKS ON DISEASES OF SPECIFIC HOSTS:

ACER:
Cristulariella depraedans (**Fig.207**)

Causes striking, white leaf spots on leaves, which fall early; usually on sycamore; fruit bodies are minute, stalked, pin-head-like structures on the underside of the leaf; an infrequent and unimportant disease.

Pleuroceras (= Ophiognomonia) pseudoplatani (**Fig.208**)
Giant leaf blotch of sycamore

Causes large, brownish, roughly circular patches on leaves, which fall early. The veins beneath blacken and bear tiny, blister-like fruit bodies; an infrequent and unimportant disease.

Fig. 207

Fig. 208

*Rhytisma acerinum (*anamorph:
Melasmia acerina) (**Fig.209**)

Tar spot of sycamore (and other maples). Causes large, striking, raised, shiny black spots on leaves which may fall slightly early. All the infections in a season arise from spores produced from the previous year's infected, overwintered leaves. An unmistakable, conspicuous and common but unimportant disease.

It is often said that the disease is rare in cities because of the fungicidal action of the air pollutant, sulphur dioxide, but there is some evidence that the explanation could equally well be the lack of overwintered leaves in cities.

AESCULUS:
Guignardia aesculi, see p.141

BETULA:
Discula (= Gloeosporium) betulina
(**Fig.210** on *B. pendula 'Laciniata'*)

Causes numerous, small, blackish spots on leaves, which yellow and fall prematurely. Fruit bodies are minute, dark blisters on dead petioles and the undersides of leaf spots. Quite common and defoliation can be severe, but it is not known to cause any lasting damage to the tree.

Fig. 210

Fig. 209

SECTION 3

CARPINUS:
Gnomoniella carpinea
*(*anamorph: *Monostichella robergei)*

Causes pale brown leaf spots of various sizes which later run together to form large blotches, often involving much of the leaf margins and tips. Fruit bodies are minute, dark blisters on the spots. Probably commoner than the infrequent reports suggest as it usually occurs rather late in the season and may be mistaken for normal autumn coloration.

CRATAEGUS, CYDONIA, MESPILUS:
Diplocarpon mespili (= Fabraea maculata.
Anamorph: *Entomosporium maculata* or
E. mespili) **(Fig.211)**

Causes numerous, small, dark reddish-brown leaf spots which coalesce to involve large parts of the leaf. Yellow patches develop on some leaves, and these fall prematurely. Spots also occur on shoots, on which the fungus overwinters, and on fruit. Fruit bodies are minute, dark blisters on the spots on the upper surface of the leaf. Infrequently reported and unimportant on ornamentals, though occasionally causes noticeable defoliation.

CRATAEGUS:
Monilinia johnsonii, see p.182

CYDONIA:
Monilinia linhartiana, see p.181
Diplocarpon mespili, see above
under Crataegus

FAGUS:
see *Quercus* below

ILEX:
Phytophthora ilicis, see p.203

JUGLANS:
Gnomonia leptostyla
*(*anamorph: *Marssonina juglandis)*
(Fig.212)

Causes numerous, large, dark brown spots on leaves and fruit, which, if heavily spotted, fall prematurely. Shoots develop dark lesions. Infected green nuts are unsuitable for pickling and kernels of nuts from badly defoliated trees are often dark and shrivelled. Fruit bodies are minute, dark blisters on the underside of the leaf spots. Widespread; common in some years, causing considerable defoliation on individual trees. This can be prevented with fungicidal sprays (as for Drepanopeziza sphaeroides on Salix – see below), but the infrequency of serious outbreaks

Fig. 211

Fig. 212

renders control unnecessary on walnuts grown primarily as ornamentals.

Very much less common is a foliage and fruit blight caused by the bacterium **Xanthomonas campestris pv. juglandis** (= X. juglandis). Infected leaves are covered in tiny spots but do not fall; shoots may be girdled and die; black blotches appear on fruits. No control is required on amenity trees.

A striking but unimportant white mould (**Microstroma juglandis,** Fig. 213) occurs occasionally on Juglans and **Carya** species: powdery or downy, white patches appear beneath the leaflets, often centred on and running along lateral veins, which remain green; on the upper surface, yellow spots and blotches coincide with the white patches. Peace (1962) considered the disease to be absent from this country but it was present in Norfolk in 1951 and has since been found in widely scattered localities as far north as Yorkshire.

POPULUS:
*Drepanopeziza populorum (*anamorph: *Marssonina populi)* **(Fig.214)**

Causes numerous, small, brown leaf spots and premature defoliation. Infection progresses from the lowest branches upwards as the season wears on. Fruit bodies are minute greyish blisters on the upper surface of the spots, exuding white spore masses in moist conditions. The disease is common and striking on Lombardy poplar, which by autumn may be left with only a few leafy branches at the very top of the tree. Severe attacks are sometimes frequent enough to result in considerable twig and branch death. Where practicable, the disease should be preventable with the fungicidal treatment given on p.163 against Anthracnose of weeping willow.

Fig. 213

Fig. 214

MALUS:
Venturia inaequalis, see p.245

MESPILUS:
Monilinia mespili, see p.181
Diplocarpon mespili, see p.164
under Crataegus

POPULUS:
Venturia populina and *V. tremulae,*
see p.245

PRUNUS:
Apiognomonia erythrostoma

A striking leaf-killing disease of Prunus avium is caused by the fungus Apiognomonia erythrostoma. During summer, brown blotches with a yellow margin develop on the leaves (Fig.216). The leaves die but do not fall (Fig.215). During winter they hang brown and withered giving the impression that the tree is dead. The fungus is confined to the leaves, however, and normal growth will resume in spring, eventually obscuring the old, dead leaves. For unexplained reasons, records of the disease became rare after the 1950s but in the early 1990s widely scattered outbreaks occurred and within a few years it was again common and widespread in Great Britain.

Fig. 215

Fig. 216 *Damage to Wild cherry caused by the fungus Apiognomonia erythrostoma*

PRUNUS:
Blumeriella jaapii
Cherry leaf spot

Round purplish spots develop in May and June, later turning brown and falling out to leave 'shot holes'. The leaf sometimes turns conspicuously orange or yellow but a green border persists around the spots or holes (Fig 217). Leaves fall early and defoliation can be severe. In damp weather a white mould grows from the underside of the spots. A troublesome orchard disease on the Continent where control is achieved with several applications, following petal fall, of the fungicides captan, dodine, di-thianon, propineb or triforin. First recorded in the U.S.A. and Europe in the 1880s but not known here until the 1960s, in nurseries, and 1993 on outplanted trees. Now often seen in the south of the country on Prunus avium. Also occurs on P. padus, P. subhirtella autumnalis, P. yedoensis and some other flowering cherries but not known on the Japanese flowering cherries of garden origin such as P. 'Kanzan' or P. 'Shirofugen'

Fig. 217

PYRUS:
Venturia pirina, see p.245

QUERCUS:

The commonest distinctive leaf spots on oak are probably those caused by the Oak leaf phylloxera (p.194) or by leaf miners in the genus Coleophora (see p.158).
A striking but infrequently recorded and unimportant leaf spotting and blotching of both Quercus and **Fagus** is caused by the fungus Apiognomonia errabunda (Fig.218). Lesions are often centred on leaf veins. Shoots may be killed.

Fig. 218

SALIX:
Drepanopeziza sphaeroides (anamorph: *Marssonina salicicola*).
Anthracnose of Weeping willow
(Figs.219-221)

Causes small, brown or purplish spots on leaves and shoots (Fig.221). Leaves often curl or become otherwise distorted (Fig.220) before falling prematurely (Fig.219). The spots may enlarge and girdle young shoots, killing the distal parts, or, as the shoot hardens, form small, black lesions which become rough, persistent cankers in which the fungus overwinters. Fruit bodies are like those for D. populorum or Populus. The disease is only of consequence on the Common or Golden weeping willow, Salix × sepulcralis nothovar chrysocoma (= S. alba 'Tristis' or S. alba 'Chrysocoma'), when repeated severe attacks can spoil the tree's weeping habit. Control can be achieved with fungicides. Effective materials are benomyl, captafol, mancozeb, maneb and quinomethionate. [Take note of current Pesticides Regulations – see p.288] Apply at bud-break then at fortnightly

Fig. 219

Fig. 220

Fig. 221

intervals until mid-summer or the on-set of hot, dry weather. Alternatively, replace persistently diseased trees with the resistant weeping Pekin willow, S. matsudana 'Pendula' or the hybrid S. × sepulcralis var. sepulcralis (see Rose, 1989a).

SALIX:
Venturia saliciperda, see p.245

SORBUS:
Venturia inaequalis, see p.245

LEOPARD MOTH   Insect: *Zeuzera pyrina*

Damage Type: Hollowing out of shoots, twigs and small branches.

Symptoms & Diagnosis: Wilting and subsequent death of young shoots, especially on Aesculus. Larger live branches and stems up to 10cm diameter may snap due to a strong wind or other physical pressure (Fig.222).

Fig. 222

Confirmation: Close examination will reveal tunnels, often deep in the wood of affected shoots and stems, with entrance/exit holes made by insect larvae (Fig.223). A tunnel may contain a larva up to about 50mm long, dull white or yellowish with a black head and body spots.

Caution: On Aesculus, do not confuse early stages with shoot damage by squirrels, which from a distance appears similar.

NOTE: Similar but larger tunnels in mature trees, often just below the bark but also deep into the wood, are caused by larvae of Cossus cossus (**the Goat moth**). These are encountered much less frequently. Full grown larvae of this species are about 85 mm long and pink to reddish-brown in colour with a black head. Repeated attacks may kill the tree.

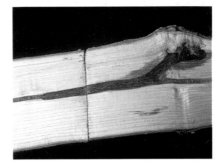

Fig. 223

Status: Widespread and moderately common in the southern half of England but rarely found in the south west or in Wales. Not found in northern England or Scotland.

Significance: May cause stem failure on feathered trees, standards or semi-mature specimens. Death of shoots on larger Aesculus is an aesthetic problem; it is not usually a cause of serious damage or concern regarding the health of the tree. Wounds often occlude on small branches, which will continue to grow. However, these are sometimes opened up, possibly by woodpeckers, and the resulting damage can appear rather unsightly (Fig.224).

Fig. 224

SECTION **3**

Host Trees: Most frequently on Aesculus, Acer and Quercus. Other hosts include Fagus, Fraxinus, Malus, Salix and Sorbus. [C. cossus is recorded from Alnus, Betula, Fraxinus, Quercus, Salix and Ulmus.]

Life Cycle: The adult moths lay eggs from late June to early August. When they hatch the small larvae mine leaves and petioles but later twigs and finally larger diameter stems. The feeding site is changed several times during the two, or even three, year development. The larvae pupate head down in the final gallery constructed.

Control: Small numbers of larvae can be killed in situ by pushing a sharp wire into the galleries. Small infested branches may be pruned out.

LIGHTNING DAMAGE

Damage Type: Physical injury; bark and foliage killing; whole tree killing.

Symptoms & Diagnosis: (Figs.225-230) Lightning causes two distinct types of damage: (i) Physical injury: A tree or a part of it is blown to pieces; or a long, narrow furrow appears on the stem where the bark and perhaps some wood has been torn out (Fig.225); or large sheets of bark are torn from the tree and deposited perhaps many yards away (Fig.226). (ii) Gradual or sudden death of all or part of single or scattered trees or of one or more groups of trees (Fig.227).

Confirmation: (i) Where the injury has clearly resulted from an explosive force, a cause other than lightning is improbable. A lightning rift typically runs in a gentle spiral from the top of a surface root or from the root collar (rarely below) to a dead or dying branch or tree top; bark fragments may be deposited far from the tree. (ii) The stem may or may not be dead but the main roots are still alive. Probing the bark of a live stem may reveal a narrow, perhaps discontinuous and upwardly tapering, longitudinal strip of dead bark disposed as described for a rift at (i) above

Fig. 225

Fig. 226

(Fig.229). Rarely, a large patch of basal stem bark is dead, above and below ground. Where several trees have been struck in a group or row, one may be shattered or bear a rift or dead strip and all will have died within a short time of each other. Killed branches may spiral up the tree, following the path of a dead bark strip (Fig.230).

Quite often, however, none of these confirmatory symptoms can be found and the diagnosis is based on circumstantial evidence (where a number of trees are affected, particularly if these are of various species and all were affected suddenly at one and the same time), together with a lack of any other satisfactory explanation.

Additional indicators: More than one tree species is involved (Fig.227).

Caution: Callused drought cracks (p.110) or the 'seams' of Bacterial wetwood (p.74) can resemble callused lightning rifts. Crown symptoms of trees struck in a group or row may develop at different rates, giving the false impression of a spreading disease condition.

Broad strips of dead bark on Fagus, sometimes bearing fungal fruit bodies, may be due to drought (see page 111).

Status: Not uncommon but often unrecognised.

Significance: Often fatal.

Trees Affected: Any species of any size

Fig. 227

(including hedges – Fig.228) may be damaged.

The Damage Process: The main and damaging electrical discharge in a lightning stroke is upwards, from ground to cloud. If this enormous surge of energy (up to 200 megawatts in a fraction of a second) passes through a tree, it is assumed that the water in the tissues along its path is instantaneously vaporized, with explosive force or fatal consequences.

Fig. 228

SECTION 3

Fig. 229

Fig. 230

Control: *Therapy:* None available.
Prevention: Trees can be protected with lightning conductors (see British Standard 6651 of 1985).

Remarks: No references to the dead bark strips ('hidden lightning scars') have been found prior to an account in the Forestry Commission Pathology and Entomology Branches' house jour-

nal 'Entopath News' No. 56 of January 1968.

Quite often, the top several feet of large Wellingtonias are dead. This damage has not been thoroughly investigated, but what little observational evidence there is indicates that lightning is the probable cause.

Further Reading: Rose, 1990.

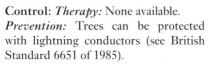

| LIME MITE | Mite: *Eotetranychus tiliarium* |

Fig. 231

Damage Type: Foliage discoloration and distortion, premature defoliation.

Symptoms & Diagnosis: In summer the leaves are grey-brown or bronzed (Fig.231), this being most noticeable on the lower crown (Fig.232) but damage may spread upwards by the end of July to affect all foliage. By August the leaves are curled and the undersides appear dirty with fine silk, mites, their moulted skins and their droppings contaminating the surface (Fig.233).

Fig. 232

Premature leaf fall often occurs by mid-August.

Additional indicator: From late July onwards throughout the autumn a silvery polythene-like sheath of silk may partly or completely cover the trunk, as well as the larger branches. This sheath is produced by numerous mites, each individual spinning a silk thread as it moves across the bark surface. The mites may be visible as a yellow dust on the bark, or in crevices, during late summer and autumn.

Caution: Similar damage but without the silk may be caused by leafhoppers (p.157).

Status: Sporadic occurrence in England, Wales and Scotland. Common some years on street trees, also on garden and parkland trees, especially in hot and dry summers.

Significance: Causes premature leaf fall in late summer; not significant regarding the overall health of affected trees.

Host Trees: Tilia spp., including T. cordata, T. × euchlora, T. × europaea, T. petiolaris and T. platyphyllos.

Life Cycle: The adult mites are yellow to darker orange-red and spider-like, up to 0.4mm long. They overwinter in bark

Fig. 233

crevices and become active when the tree flushes. They feed and breed on the underside of the leaves in summer. When the leaves become unsuitable for feeding (late July onwards if the infestation is heavy) the mites move onto the trunk and branches in large numbers.

Control: Not necessary. It is unusual for infestations to occur on the same tree in consecutive years.

SECTION 3

LIME- OR CHALK-INDUCED CHLOROSIS
also other deficiency diseases

Cause: *Iron and/or manganese deficiency; deficiencies of other soil nutrients*

Damage Type: Impaired chlorophyll synthesis resulting in foliage yellowing and dieback.

Symptoms & Diagnosis: *Broadleaves:* The leaf tissues between the main veins are yellow while the veins remain green (Fig.234); *Conifers:* needles or, in cypresses, whole fronds are uniformly yellow (in pines, some needles may be yellow only at their base). Perhaps in addition, some degree of leaf browning, leaf cast and dieback is evident (Fig.235).

Additional indicators: Needles and leaves are smaller than normal; damage is most severe in the upper and younger parts of the tree and varies in severity from branch to branch; affected trees stand next to healthy ones; the condition has spread and intensified year by year with needles and leaf margins browning, premature defoliation, more and more trees showing symptoms, dieback setting in and trees dying. The main roots of dying trees are found to be the last part to die.

Confirmation: Field diagnosis depends on finding these symptoms on susceptible tree species on an alkaline site in the absence of other explanations. The precise nature of the deficiency can only be determined by comparing the iron and manganese content of affected leaves with leaves from healthy trees.

Caution: Interveinal chlorosis and dieback can be induced by other elemental deficiencies, and by some toxic chemicals. On chalk, for example, such symptoms may be due to potassium deficiency. To distinguish, a chemical analysis of the foliage in autumn is necessary. See also chemical damage, p.93.

Status: Common in susceptible plants on calcareous soils.

Fig. 234

Fig. 235

Significance: A principal determinant in the choice of species for alkaline sites. The disorder can be very damaging where the wrong choice has been made. Susceptible and Tolerant Species: See lists below.

Reasons for the Disorder: The chemistry of soils containing much calcium renders iron and manganese (elements essential in the synthesis of chlorophyll) unutilisable by certain (calcifuge) plants. The precise reasons for this are complex, appear to differ on different soils and are incompletely understood. Some (calcicole) plants are, however, inherently able to grow normally or relatively healthily on such soils.

Control: *Therapy:* To allow the development of an acidic humus layer in which adventitious rooting can occur, the soil around affected trees should not be disturbed. The addition from time to time of a lime-free organic mulch or thin layers of an acidic, porous soil can accelerate this slow process. Repeated treatment of soil with chelated iron (a form soluble even in alkaline soils) or implantation of stems with chelated iron, iron salts or manganese sulphate can alleviate symptoms. Often, however, affected trees are best replaced with lime-tolerant species.
Prevention: Do not plant calcifuge species on calcareous sites.

Remarks: Often, lime-induced chlorosis appears only after some years of healthy growth. Symptoms are intensified in hot, dry summers.

TREE GENERA and SPECIES PARTICULARLY TOLERANT or INTOLERANT of CALCAREOUS SOILS

Tolerant		Intolerant	
Abies cephalonica,	Parrotia	Abies (some spp)	Nyssa
A. concolor var.	Phellodendron	Acacia	Oxydendrum
lowiana	Phillyrea	Amelanchier	Picea (most spp)
Acer	Picea omorika	Castanea	Pinus (most spp)
Aesculus	P. pungens & vars.	Cornus	Prunus laurocerasus
Alnus cordata	Pinus cembra	(most tree spp)	Pseudolarix
Betula	P. leucodermis	Embothrium	Quercus (many spp)
Buxus	P. mugo	Eucryphia	Sassafras
Carpinus	Pinus nigra & vars.	Halesia	Sciadopytis
Cedrus	(especially Austrian	Liquidambar	Stuartia
Cephalotaxus	pine)	Magnolia	Styrax
Cercis	Podocarpus	(many spp)	Taxodium
Chamaecyparis	Populus alba	Nothofagus	Tsuga heterophylla
Cotoneaster	Prunus (most spp)		
× Cupressocyparis	Pterocarya		
Crataegus	Pyrus		
Cupressus	Quercus cerris		
Fagus	Sequoiadendron		
Fraxinus	Sorbus (Aria		
Ilex (most spp)	Section)		
Juniperus	Taxus		
Maclura	Tetracentron		
Magnolia kobus	Tetracentron		
Malus	Thujopsis		
Morus	Torreya		
Myrtus	Ulmus		

NOTE: There is general agreement amongst authors over which species should be regarded as intolerant of calcareous conditions; there is less agreement over lists of tolerant species. Some of those given here as tolerant (e.g. Fagus, Larix, × Cupressocyparis – Fig.235) will become chlorotic or even die back on the limiest of soils.

Most species not listed here are neither particularly tolerant nor intolerant but for a few of these information is lacking.

SECTION 3

OTHER SOIL NUTRIENT DEFICIENCIES

Symptoms, Diagnosis and Control: The most noticeable indications of soil nutrient deficiencies are the discoloration, usually yellowing or purpling, of foliage (often interveinal and marginal) and, in extremis, the death of the discoloured parts. The discoloration may be confined to either the older or to the younger leaves or needles. Growth may be stunted, leaves undersized, and dieback may occur. Although deficiency diseases in specific orchard and forest trees on certain site types have been characterized and well documented, the diagnosis and therefore the correction of deficiency diseases in amenity trees requires specialist knowledge and chemical analysis. Treatment should be based on soil analysis and tailored to the case in point; the unregulated addition of fertilizers can lead to further nutrient imbalance or deficiencies and further problems. Help may be obtained from ADAS or (for RHS members only) from the Royal Horticultural Society, or from the Forestry Commission's research stations (see p. XX for addresses). *Caution.* Except on very alkaline soils (see lime-induced chlorosis above), very acidic soils such as peats, or on very sandy or gravelly soils lacking organic matter, severe deficiency symptoms in trees are far more likely to indicate an inability of the roots to take up available nutrients than that the soil is lacking them. This can be the effect of drought (p. 112), waterlogging (p. 266), disease or, in recently planted trees, severe competition from a grass sward (p. 7). Other problems such as the uptake of toxic chemicals (p. 93) or certain insect feeding can also give rise to similar symptoms.

MECHANICAL DAMAGE

Causes: *Biting insects & mammals, machinery, tree ties, flying objects, snow & ice, wind, and similar agents; grafting*

Fig. 236

Fig. 237

Damage Type: Removal, severance, crushing, tearing, abrasion or piercing of tissues. Also, imperfect graft unions, root asphyxiation.

Symptoms & Diagnosis: (Figs.236–250) Usually, that the damage is mechanical is evident on even fairly cursory examination, though the precise cause may not be: **pieces of tissue are missing, strips of bark hang partly stripped from the stem, branches or shoots lie broken on the ground, a branch has been abraded.** The cause must then be deduced as described in Section 1.

Disappearance of parts of or whole leaves or needles: consider the possibility of caterpillar or weevil or beetle damage and search for the insects themselves or signs of them such as cast skins, frass or silken threads. See 'Remarks' on p. 181.

General crown deterioration (perhaps in conjunction with **sprouting from the root–stock**) can indicate the constriction of the stem by a tree tie, girdling root (Fig.239 showing tree snapped below its strangulated stem base) or tree guard (also check below ground for its remains); basal bark removal (common causes are mowers, strimmers, field voles, rabbits, horses, cattle, pigs); or graft incompatibility (particularly where the scion is a different species from the rootstock – compare the foliage from one with the other. See also Black line disease of walnut, p.80)

Crown deterioration (Fig. 238) (perhaps succeeded after a year or two by the development of epicormic shoots on trunk and limbs and ultimately the formation of a new crown within the original one that died back) can also arise if the roots have been starved of air (oxygen) as the result of soil compaction (most commonly by the repeated passage of heavy machinery over the rooting area, as during building operations); or where the soil level has been raised above them (indicated by the absence of visible buttress roots); or where the soil surface has been sealed by an impermeable covering such as concrete.

The death of all foliage on one or several branches in an otherwise

Fig. 239

Fig. 238

Fig. 240

3

SECTION

healthy crown is often the result of the stripping or gnawing away of bark on those branches by animals, usually squirrels (particularly common on sycamore - Fig.241 - and beech - Fig.242), bank voles (particularly in yew hedges - Fig.240) or occasionally edible dormice. Similar girdling bark injury on branchlets and twigs is likely to be due to beetle or weevil feeding. Neglected, strangulating label ties (Figs.236, 237), clothes lines and the like give identical foliar symptoms. Where no branch constrictions, or missing bark (or dead bark - see Canker diseases in index and Section 2) can be found, carefully remove bark from symptomatic branches to check for insect tunnels (feeding or breeding galleries) in the inner bark and cambium (see in particular Agrillus damage on Crataegus, p. 33). Low branches may be killed and perhaps broken where bark has been abraded by deer fraying.

Fig. 241

A lack of flowers, shoots or fruit may be the result of bird (Figs.243,244) or insect damage to these parts or, earlier, to buds (see also Ash bud moth, p.68).

Cleanly severed, live leading shoots of the pines and spruces have probably been snapped by wind or when perched on by birds (see also Pine shoot moth, p.211).

Horizontal rows of small, diamond-shaped wounds on tree stems are usually the work of woodpeckers (Figs.245 and 396).

Sometimes in summer, limbs with no apparent defect snap and fall in calm weather – a phenomenon known as **'summer branch drop'**. Factors put forward to account for cases where fungal decay and prior injuries can be ruled out include: (i) reduction in wood strength by Bacterial wetwood infection (see p.74); (ii) an increase in branch water content, and therefore weight, resulting from a reduction in transpiration during periods of calm, humid weather; (iii) a weakening of wood-cell cementation due to increased ethylene

Fig. 242

production by the tree in conditions of high humidity and temperature and low transpiration; and (iv) an increase in weight due to the production of fruit.

Young trees may snap where wood decay has developed behind wounds inflicted by mowing or weed-control machinery, and such wounds often lead to undesirable **sprouting from the rootstock** (Fig.246).

Fig. 243

Fig. 245

SECTION 3

Fig. 244

Fig. 246

A cluster of weeping spots on trunk or limbs may indicate injury by beetles – or shot-gun pellets (Figs.249, and 247 which shows the pellets embedded in the wood); **brown spots and lines on leaves** may mark the perforations and scratches caused by the thrashing of a spiney tree in the wind; **tattered leaves or broken needles and small wounds along one side of young twigs** may be the result of a hail storm;

a **narrow, horizontal swelling on a stem** may result from wind-flexing violent enough to crack but not to snap the stem (Fig.248). **Scattered, dead, drooping Horse chestnut shoots** have probably been damaged by squirrels (Fig.250) or Leopard moth larvae (p.169). In all such cases it is necessary to be particularly alert to the implications of, for example, directional damage, to scrutinize the affected parts with a hand

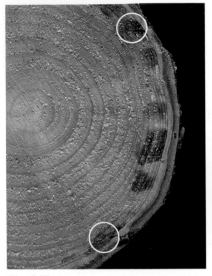

Fig. 247

Fig. 249

Fig. 248

Fig. 250

lens, to be ready to dissect them to seek the causal agent, and to date the injury.

Status: Damage from biting insects (caterpillars, beetles, weevils), squirrels and birds is common and widespread. On occasion, ice, snow and wind breakage and windthrow are widespread. Other types are occasional and localised.

Significance: Recovery from wound-ing is often spontaneous and complete. However, wounds offer entry points for wood decay and pathogenic organisms which are often more damaging than the wounds themselves, while severe or repeated injury can itself kill trees or render them too unsightly to retain. Where roots are deprived of oxygen by being covered or compacted, dieback can be expected if more than half the root spread is involved. The likelihood and severity of damage further increases with the clay content (impermeability) of the compacted or added soil. Recovery depends on these factors and on the tree species. Beech, for example, responds very poorly to a changed rooting environment whereas oak often eventually recovers, though this may take years.

Control: *Therapy:* To facilitate callusing, **trim back splintered wood, torn bark or broken branches** (see Section 5, Decay). Guard against infection from Silver leaf (p.252) or Bacterial canker (p.70) as necessary.

It may be possible to save **trees recently girdled by bark stripping** by means of bridge- or in- grafting or bark implants (see Harris 1992 or consult a horticulturist or orchardist). Meanwhile, prevent the exposed wood from drying out by covering it in polythene sheeting.

A broken leader in species which do not resprout readily from dormant or adventitious buds (many conifers, for example) may need replacement by the careful training in of a side branch.

Multiple leaders resulting from such damage on other species may need singling.

Experiments to date suggest that the loosening of **compacted soil** by means of compressed air could be beneficial to a tree in a light soil but is unlikely to help if the soil is heavy.

Girdling constrictions, such as tree ties and girdling roots, should be removed, or severed at several points.

Preventive measures are usually self evident or not available. For the prevention of mower and strimmer damage, see Patch and Denyer, 1992.

Remarks: The identification, significance and control of most mammal and biting insect damage is outside the scope of this book. Useful reference books include Alford(1991) and Bevan (1987)

N.B. The damage from paving over roots may be exacerbated by the preparatory soil levelling and compaction and, sometimes, by the prior application of persistent, soil-acting weedkillers (as in the construction of hard tennis courts).

SECTION 3

MONILIA DISEASES OF QUINCE AND MEDLAR	Fungi: On Cydonia: *Monilinia linhartiana* (= *Sclerotinia cydoniae*) Anamorph: *Monilia necans* On Mespilus: *Monilinia mespili* (= *Sclerotinia mespili*) Anamorph: *Monilia mespili*

Fig. 251

Damage Type: Leaf and fruit killing.

Symptoms & Diagnosis: In spring, scattered leaves are marked with dark brown or blackish blotches, the lesions often centred on the midrib and extending down the petiole (Fig.252). Affected leaves remain for some time hanging dead on the tree.

Later in the season, dry 'mummified' fruits hang dead on the tree. In quinces (it is not clear from the literature if this applies to medlars, but it may well), scattered dead, current shoots may be present.

Confirmation: Some affected leaves and fruits, especially during or after damp weather, are covered in a whitish, sweet-smelling mould (Fig.251, the Monilia sporing stage of the fungus).

Status: Uncertain. Probably both diseases occur occasionally wherever the host plants are grown.

Significance: Little more than curiosities, though potentially disfiguring and damaging to quince or medlar fruit crops should weather conditions favour a severe outbreak.

Host Trees: Cydonia oblonga and Mespilus germanica.

Infection & Development: Although the two fungi are each confined to their respective host trees, the diseases they cause are probably virtually identical so are treated as one here.

Flowers and young, unfolding leaves are infected in spring, mostly from spores produced from twigs and dead fruits in which the fungus has overwintered, occasionally from fruit bodies on the fallen, infected fruits. Parts or all of the leaf blade and often the petiole are killed. The fungus may grow into the shoot from the petiole, girdle it and so cause the distal part to die. From the flower, the fungus spreads into the young fruitlet, soon permeating it,

Fig. 252

arresting its development and finally killing it. The infected fruit remains hanging in a mummified state on the tree for a while before dropping.

Control: In ornamentals, the disease is probably too rarely severely damaging to warrant control, but if it proves troublesome on trees grown for their fruit, control measures advocated for the similar disease on Prunus (Blossom wilt and spur blight, p.85) would be effective.

Remarks: The disease was studied at the end of the 19th century, in France.

MONILIA LEAF BLIGHT OF HAWTHORN	Fungus: *Monilinia johnsonii* (= *Sclerotinia crataegi*) Anamorph: *Monilia crataegi*

Damage Type: Leaf killing, fruit killing.

Symptoms & Diagnosis: Dark brown or blackish, dead leaves hang down limply, often singly and inconspicuously scattered among mostly healthy foliage (Fig.253); other limp leaves show similarly coloured, irregular blotches, often centred on the midrib and running into the dead petiole. The fungus also infects the haws, on which dark lesions would presumably at first

Fig. 253

Fig. 254

develop. Later in the season, dry 'mummified' fruits are found on the tree.
Confirmation: Some affected leaves, especially in or after damp weather, are covered in a greyish, characteristically sweet-smelling mould (Fig.254).

Status: Uncertain. Known from scattered localities in England and southern Scotland but probably widespread in Great Britain and overlooked. One authority says it is very common in S.E. England. It rarely causes enough damage to draw attention to itself.

Significance: There seems to have been only one recorded serious outbreak of the disease in Great Britain – in Kent and East England in 1937. Otherwise, the damage involved is probably inconsequential, rarely even disfiguring the trees, but the rather unusual symptoms do puzzle the observer.

Host Trees: Crataegus monogyna, C. laevigata and, according to one authority, on many other species of Crataegus. Also on × Crataemespilus.

Infection & Development: The fungus overwinters on fallen mummified fruit. On these in spring are borne the short-lived, perfect (Monilinia) fruit bodies of the fungus. The spores from these are carried in the air to infect developing leaves. The fungus grows through and kills the leaf; often the area around the midrib and the petiole dies first. The leaf remains hanging dead on the tree for some time and, in suitably moist conditions, becomes covered in a grey, sweet-smelling mould (Fig.254) consisting of large numbers of asexually produced (Monilia) spores, possibly attractive to and so spread by insects. These infect the young fruits which become permeated and mummified by the fungus and eventually fall.

Control: None required.

Remarks: Known in the USA and Germany in the 19th century; the Monilinia state was first noted in 1905, in Germany. The disease was first recorded in this country in 1926, from Kent.

SECTION 3

MOSSY WILLOW-CATKIN GALLS

Cause: *Possibly a virus*

Fig. 255

Fig. 256

Fig. 257

Damage Type: Virescence, i.e. the transformation of flowers (catkins) into bunches of vegetative shoots (small witches' brooms).

Symptoms & Diagnosis: At any time of year, dense clusters of shoots (galls or small witches' brooms), roughly spherical in outline and up to about 8 cm in diameter (Fig.255) or elongated and pendulous, are apparent scattered through the tree's crown, perhaps in large numbers (Fig.256). In summer, both live and leafy and dead galls may be present; soon after flowering time, dead, fully formed galls may be seen but live galls are only beginning to develop and appear less conspicuously as much-enlarged catkins (Fig.257 arrow).

Confirmation: On close examination, the galls are seen to be modified catkins: the axis (rachis) of the female catkin is thickened; some or all of the immature fruitlets have been transformed into distorted shoots bearing small leaf-like structures while a proportion or all of any remaining fruitlets are enlarged and more or less deformed (Fig.257).

A similar change in male catkins is also said to occur, together with the formation of witches' brooms and fasciation involving vegetative shoots, but we have not observed this.

Nothing else on Salix resembles these galls.

Status: Uncertain, though locally common in the southern half of England.

Significance: An attractive curiosity when alive and in leaf, though if numerous may be regarded by some as unsightly when dead.

Host Trees: Certainly on Salix alba and S. fragilis in Great Britain. Also reported on S. caprea but few of the very few records traced state the species of willow involved.

Infection & Development: Not fully elucidated. The abnormalities develop during the tree's growth in spring and galls die within 12 months. When fully formed, the galls are home to large numbers of mites and various insects.

Control: None required.

Remarks: Despite the conspicuous nature of these galls and their description at least as early as 1898, they are rarely mentioned in the literature. They have been attributed to mites (in particular Phytoptus (= Eriophyes) triradiatus), but workers in the 1940s and 50s pointed out that these merely took up residence in the galls after their formation and were not their cause. In 1975, French workers demonstrated the presence in affected tissues of virus-like particles which were absent from normal plants, but whether these are indeed the causal agent has not yet been determined.

NECTRIA CANKER

Fungi: *Nectria galligena and N. ditissima*
Anamorphs: *Cylindrocarpon heteronemum and C. willkommii*

SECTION 3

Fig. 258

Fig. 259

Damage Type: Bark killing.

Symptoms & Diagnosis: (i) In the growing season, scattered twigs or shoots are seen to be leafless or bearing only wilted or dead leaves (Fig.258). (ii) Large or small, perennial 'target cankers', centred on a dead twig, bud or injury, are evident on stem or branches (Figs.259, 261).

Confirmation: If the bark of an affected twig or shoot is alive, either a ring of dead bark will encircle it at its base, perhaps as a sunken canker, or part way along, probably at the base of a year's extension growth (Fig.260). On dead twigs, these old girdles may still show as patches of flaking or cracking bark. Small white pustules (conidia) or red, pin-head like fruit bodies, much like those of Nectria coccinea (Fig.45), may be present on the dead bark or at margins of cankers.

Status: Widespread and very common.

Significance: One of the most serious of orchard apple and pear diseases. It is sometimes conspicuous but rarely of any consequence on ornamentals.

Fig. 260

Fig. 261

Host Trees: Malus, Pyrus, Sorbus aucuparia, Fagus sylvatica, Fraxinus (cf Bacterial canker of ash on p.69). Also probably Prunus, Acer, Ilex: see Remarks below.

Resistant Trees: The disease occurs occasionally but is of little consequence on many broadleaves in addition to those mentioned above. Conifers are immune.

Infection & Development: Spores (conidia or ascospores) carried in wind or rainwater infect bark through injuries or natural wounds, most often leaf scars in autumn and bud scale scars in spring. In the tree's dormant season, the fungus grows through and kills a small patch of bark, perhaps enough to girdle a shoot or small twig. On larger members, callus tissue forms at the margins of lesions in summer, only to be killed back again in the following winter. Repetition of this callusing and killing results in a sunken 'target canker' which may continue to enlarge for many years.

Control: Not required in ornamentals, though for appearances' sake, dead twigs can be removed. Cankers on larger members rarely girdle and can safely be left. Control on fruit trees is by fungicides, the use of resistant varieties and orchard management.

Remarks: The disease was known in Europe before 1866. Nectria galligena was originally described from galls on Salix purpurea. The large target cankers seen on Flowering cherries, sycamore, holly and occasionally on other broadleaves are probably also caused by this disease.
A disease of Morus species (Fig.261) which is indistinguishable in the field from Nectria canker is caused by the fungus Gibberella baccata (anamorph: Fusarium lateritium). The few accounts in the literature suggest that it does not warrant control measures beyond the removal of dead and dying twigs.

NEEDLE CAST DISEASES
Fungi: *Various species*

Damage Type: Needle discoloration; needle death; defoliation.

Symptoms & Diagnosis: (Figs.262-270) At any time of year, a tree's crown appears thin due to the loss of needles; OR many needles are discoloured (pale green, yellow, orange or brown), either uniformly or variously spotted, mottled or banded; the discoloured needles may or may not be falling when observed.

Additional indicators: Affected needles are all of the same age (i.e. the current year's, or the previous year's or an earlier year's). Unaffected needles are normal in size and shape.

Confirmation: Fruit bodies or spore masses of the causal fungus are present on affected leaves. (Those of some species are distinctive enough to be identifiable in the field; others require laboratory examination – see Remarks on diseases of specific hosts below).

Caution: Similar needle discoloration and killing may result from pest infestations, toxic chemicals, nutritional disorders, drought, winter cold or spring frost and some of these also cause premature needle loss.

Status: Severe outbreaks of these diseases in amenity trees are uncommon as they require the rather unusual coincidence of several factors. Local climatic conditions are of overriding importance, a suitably susceptible tree and a local spore source is essential, and other foliar micro-organisms can be influential.

Significance: Severe defoliation in several successive years can lead to dieback and even death, but such repeated outbreaks are rare in amenity trees. Usually after one or two years of fresh growth the thinnest of crowns is well furnished with needles again. Exceptionally, the stem of a badly defoliated pine may be attacked and perhaps killed by the Pine shoot beetle (p.209).

Host Trees and Infection & Development are given below.
Control is not required.

REMARKS ON NEEDLE CAST DISEASES OF SPECIFIC HOSTS

LARIX:
Meria laricis (**Fig.262**)

Occasionally this causes noticeable defoliation of Larix decidua in the wetter parts of the country. Needles die from the tip back. Those completely killed eventually fall; those whose bases remain green do not. The needles at the tips of shoots remain healthy. The fungus overwinters in infected needles on the tree and in spring spores from these infect the new needles. The fruit bodies are not identifiable in the field.

The disease was first described from France in 1895 and from Great Britain in 1919.

Fig. 262

PICEA:
Chrysomyxa needle rust (**Fig.263**)

Two species of Chrysomyxa occasionally cause disconcertingly bright and conspicuous needle discoloration on spruces, most commonly in Scotland on Picea abies (Fig.263) and P. pungens.

At any time of year, crowns may appear thin from the premature loss of needles, or bright orange or yellow transverse bands are evident on needles on the tree.

Chrysomyxa abietis (Fig.263) overwinters in the discoloured, infected needles on the tree. Spores (teliospores) released from orange, elongated blister-like pustules on the underside of these in spring infect newly developing needles; the spore-bearing needles then fall.

Chrysomyxa rhododendri overwinters in reddish blotches on leaves of rhododendron. In spring, airborne spores (teliospores) produced from these infect developing spruce needles. In late summer, delicate white outgrowths protrude from the undersides of the yellowed needles (Fig.264), rupturing to release powdery orange aeciospores, but these cannot reinfect spruce needles; instead they infect rhododendron leaves. The infected needles then fall. Both these fungi were described from the Continent in the first half of the 19th century and first noted here, in Scotland, just before the first World War.

Fig. 263

Fig. 264

PINUS:
Lophodermium seditiosum (**Fig.265**)

This indigenous species is the commonest and most widespread of the needlecast disease fungi on pine. It occurs on many 2-, 3- and 5-needled pines but Pinus sylvestris is the only species on which noticeable defoliation is likely to be encountered.

Airborne spores give rise to infections on current needles, mostly in periods of high humidity in late summer and autumn. Small yellow spots develop at first, later turning brown. During winter the whole needle turns a characteristically pinkish brown colour. Browning may be noticeably one-sided on the shoot. Defoliation occurs in late winter and spring, so that in the summer following infection a tree may bear virtually none of the previous year's needles.

Fruit bodies do not appear until some time after needles have fallen so are not useful for field diagnosis. The fungus is also a common and harmless inhabitant of living pine needles. If these are killed prematurely, as for instance when a tree is felled or live branches are cut off, L. seditiosum may fruit abundantly on them and lead to an outbreak of disease in trees nearby.

PINUS:
Lophodermella sulcigena (**Fig.266**)

This widespread fungus occasionally causes dramatic browning of the current needles and sometimes premature fall of second-year needles of 2-needled pines, notably, among the commoner amenity species, Pinus nigra and its varieties, and P. sylvestris.

In wet summers, airborne spores infect the unhardened tissue at the base of growing needles. Because needles grow from their base, the point of infection is carried away from the needle sheath until growth ceases. The fungus kills a band of tissue around the needle which dies back to this point. By the end of summer, therefore, large numbers of the current needles may be brown though the needle bases typically remain green (Fig.266 arrow).

Spores are produced from the infected needles in the following summer, after which the needles fall. Often, though,

Fig. 265

Fig. 266

SECTION 3

other fungi colonize the dead tissues and prevent L. sulcigena from fruiting and the needles from falling.

The fruit bodies are not valuable for field diagnosis. The disease was first described in 1892, from Norway. The first British report was on P. sylvestris in Scotland in 1920.

PINUS:
Cyclaneusma (= Naemacyclus) minus
(Fig.267)

This fungus is occasionally responsible for the summer yellowing and premature fall of needles of any age from various 2-, 3- or 5-needled pines, notably Pinus radiata (Fig.267 with healthy Pinus sylvestris). Infection from airborne spores produced on fallen needles takes place at various times of year. Up to a year later, infected needles turn yellow, often with transverse brown bands, and fall, usually in autumn. In humid weather (or if the needles are kept damp in a closed box for a week or so), the distinctive fruit bodies develop on these brown bands, opening by two flaps like a double trapdoor (Fig.268). The history of the disease is confused as only in the 1970s did mycologists distinguish the causal fungus from other related species.

Fig. 267

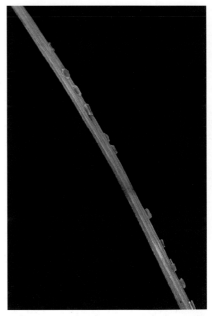

Fig. 268

PSEUDOTSUGA:
Phaeocryptopus gäumannii (**Fig.269**)

Crowns of Pseudotsuga menziesii in the west and north of the country sometimes look yellowish and thin due to needle infection by this fungus.

Most infections occur on current needles during early summer. Infected needles soon produce sporulating fruit bodies, even while still green, and continue to do so, gradually becoming mottled yellow then brown until, up to 3 seasons later, they fall. The minute, black fruit bodies protrude through the stomata on the underside of the needle, appearing therefore as two narrow, sooty bands composed of tiny dots (Fig.269).

This North American fungus was first noted in Europe and Britain in the 1920s.

PSEUDOTSUGA:
Rhabdocline pseudotsugae (**Fig.270**)

From autumn through to spring, following a cool, wet summer, the crowns of Pseudotsuga menziesii may turn conspicuously brown due to infection of current needles by this fungus; in summer, crowns appear thin when these needles, now the previous year's, have been shed. Spores are released from browned needles on the tree from May to July and infect the currently developing needles. Purplish brown patches and bands develop, giving the foliage a strikingly speckled appearance. In the following spring, long blisters form beneath the needles on the lesions, splitting longitudinally to release the spores (Fig.270). This North American fungus was first noted in Europe and Britain in the 1920s.

SECTION 3

Fig. 269

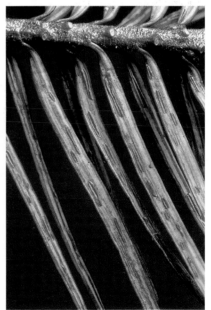

Fig. 270

OAK DIEBACK

Cause: *A series of adverse factors, biotic and/or abiotic*

Fig. 271

Fig. 272

Damage type: A complex disease or 'decline'

Symptoms and Diagnosis: *Early symptoms*: Leaves over all or much of the crown are smaller than normal, pale green or yellow and perhaps sparser than usual (Fig. 271). *Later symptoms*: As earlier but dead twigs and branches are also present scattered through crown. *Finally*: All or most limbs have died back and the remaining live foliage is pale or yellow and scattered sparsely among many dead twigs. Recovering trees are typically 'stag-headed': the ends of dead limbs, bearing few fine twigs, stick out prominently from the green, foliated part of the crown which consists of healthy young branches (epicormic shoots) growing from the trunk and lower parts of the main limbs (Fig. 272).

Confirmation: The roots are alive, even on trees which have died back severely.

Additional indicators: Patches of dying or dead bark, perhaps extensive and girdling, perhaps indicated by dark exudates ('tarry spots'), are present on the lower bole (Fig. 273). If this bark is levered off, a pattern of very sinuous grooves 3-4 mm wide, perhaps forming a complex network, is evident on the inner bark and outer wood surfaces (Fig. 274). In some grooves, larvae of the buprestid beetle, Agrilus pannonicus, may be present. These are long, much flattened, whitish, slightly translucent, legless grubs with a brown head. The exit holes are D-shaped and measure 3 to 3.5mm along the straight edge.

Caution: As explained below, different combinations of adverse factors can give rise to a similar decline in the health of the tree; future 'oak dieback' may not involve the buprestid beetle at all.

Status: The condition was widespread in eastern and central Great Britain during the 1980s and 1990s (see also Remarks below).

Host trees: Quercus robur and some hybrids between Q. robur and Q. petraea. On the Continent, Q. petraea has also suffered.

Reason for the damage: Still not fully elucidated, but these episodes of dieback here and abroad appear to be classic 'declines' (see p. 8) where the initiating stress and the subsequent precipitating factors have varied from time to time and place to place but combine to produce much the same symptoms. The early 1990s episode in this country appears to have begun with the severe droughts of 1989 and 1990 and subse-quent deaths of drought-stressed trees brought about largely by the buprestid beetle, Agrilus pannonicus.

Control: None known.

Remarks: Comparable episodes of dieback have occurred in this country and on the Continent at various times during the 20th century and up to the present. These have involved variously caterpillar defoliations, drought, oak mildew outbreaks, root killing by the fungi Armillaria or Collybia fusipes, wind, low winter temperatures and bark killing by the beetle Agrilus pannoni-cus. Research interest at the time of writing centred on the possible involve-ment of Phytophthora root-killing fungi.

Further reading: Gibbs and Greig, 1997, Gibbs 1999.

Fig. 273

Fig. 274

SECTION 3

OAK LEAF PHYLLOXERA

Insect: *Phylloxera glabra*

Damage Type: Browning of foliage.

Symptoms & Diagnosis: Extensive browning of leaves in August, sometimes restricted to one or more boughs, most often on large old trees (Fig.275). Damage first becomes visible in late June or early July as yellowish spots, 2mm diameter, on top of the older leaves. These spots then enlarge, spread and coalesce as the season progresses (Fig.276).
Confirmation: Close examination of the underside of affected leaves in summer shows circular marks, discernible even when most of the leaf is discoloured and brown (Fig.277).

Status: Widespread, sometimes common in the southern half of Britain.

Significance: Extensive browning of the foliage in late summer is of little importance to the health of the tree. Often only on a single tree within a group; sometimes only one or two of the lower branches are affected, rather than the whole crown.

Host Tree: Quercus robur.

Life Cycle: The eggs overwinter in bark crevices and the nymphs soon hatch and feed on the underside of young leaves. When mature they lay transparent eggs in a circle around themselves (Fig.278, and see Confirmation above) from which the next generation of larvae spread out and, as they mature, moult into a winged form and leave the feeding sites.

Fig. 275

Fig. 276

Control: Not normally considered; the exception may be young trees and nursery stock which could be sprayed with a suitable contact aphicide in May. [Take note of current Pesticides Regulations – see p.288] Ensure that the underside of the leaves is wetted.

Remarks: Phylloxerids are closely related to the true aphids, but differ morphologically in their wing venation, the number of antennal segments and the lack of siphunculi (cornicles) on the abdomen (see also p.274). All phylloxerids feed exclusively on dicotyledonous plants.

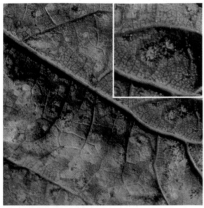

Fig. 277

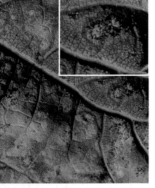

Fig. 278

OAK SLUGWORM (OAK SLUG SAWFLY)

Insect: *Caliroa annulipes*

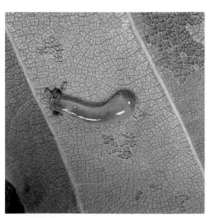

Fig. 279

Damage Type: Removal of leaf epidermis and partial leaf killing.

Symptoms & Diagnosis: In May and June brown patches appear on the leaves; further damage occurs in August (Fig.280). Close examination reveals that the lower epidermis has been removed (Fig.279), leaving a network of veins against necrotic areas of the upper epidermis.

Confirmation: When viewed against the light the brown patches appear translucent. Check for moulted skins of sawfly larvae adhering to the leaf surface with their front legs splayed out. Damage is rarely noticed until after the insects have departed; however in May and June or August slug-like pale yellowish-green, shiny larvae up to 10mm long and with dark heads may be found feeding (Fig.279).

Status: Widespread and common.

Significance: Of no consequence on large trees but can be unsightly on smaller ornamentals.

Host Trees: Most frequent on Tilia; also on Betula, Crataegus, Fagus, Prunus and Quercus.

Life Cycle: Two generations of larvae per year feeding in early summer and again in August is normal. Exceptionally there is a third generation in the autumn when conditions are favourable. The adult sawflies lay their eggs into slits in the underside of the leaves. The larvae feed in groups before dropping to the soil to pupate in cocoons. The adult female emerges from this and lays eggs into slits which she cuts with her saw-like ovipositor in the underside of the leaves.

Control: If necessary, on small trees, infested leaves with larvae present can be handpicked and destroyed.

Remarks: C. cerasi causes similar damage to the upper epidermis, mainly on rosaceous trees (see p.198.) C. cinxia causes similar damage to the lower epidermis of Quercus in the autumn, when the affected leaves appear similarly translucent but are a dirty white colour rather than brown.

Fig. 280

PEACH LEAF CURL
and other
TAPHRINA LEAF BLISTERS (GALLS)

Fungi: *Taphrina species*

Fig. 281

Fig. 282

Fig. 283

Damage Type: Deformity and killing of soft tissues.

Symptoms & Diagnosis: (Figs.281-283) Parts of leaves or whole leaves are cockled or blistered and thickened; distorted areas are discoloured: in Alnus (Fig.281) at first pale green then ash grey; in Populus (Fig.283) bright yellow on the under (concave) side of the blister, though not until after the blister is well formed; in Prunus (Fig.282) pale green then yellow, then sometimes red. Finally, the distorted parts brown or blacken and die. In Alnus and Prunus (but not Populus) the dead leaves fall. In Prunus: many dead twigs may be present from earlier infections; severe defoliation may occur and shoots die; shoots may be thickened and distorted; flowers may be infected, shrivel and fall.
Confirmation: In Populus, look for the yellow coloration (asci) beneath the blisters (Fig.283); in Alnus the severe malformation cannot be confused with anything else.
Additional indicators: Likely to be prevalent after a cool, wet spring.
Caution: In Populus, this can be mistaken for the rust disease Melampsora (p.241) on account of the conspicuous yellow colour. In Prunus, leaf aphids can severely distort leaves (p.99) but do not cause bright discoloration.

Status: Widespread. Common on Populus and Prunus, occasional on Alnus.

Significance: Peach leaf curl is a severely debilitating and disfiguring disease and can greatly reduce fruit crops; repeated defoliations can kill small trees. On Alnus and Populus, the diseases are little more than a curiosity.

SECTION 3

Host Trees:
Alnus (Taphrina tosquinetii): A. glutinosa, A. incana.
Populus (T. populina = T. aurea): most
species, commonest on P. × eurameri-
cana hybrids.
Prunus (T. deformans) 'Peach leaf curl':
P. dulcis (almond) and P. persica
(peach) very susceptible, P. armeniaca
(apricot) less so.

Resistant Trees: Alnus: rare or not
recorded on species not listed above.
Populus: probably not on P. tremula.
Prunus: species not listed above.

Infection & Development: The fungus overwinters as spores on bark and
between bud scales. In spring, in wet
weather, the spores immediately infect
the new leaves and shoots as they
emerge. As they expand, they become
distorted as a result of the secretion of
growth-regulating substances by the
fungus. Spore-bearing asci develop as a
whitish bloom covering the upper surface of the distorted leaf tissues of
Prunus and Alnus or as a bright yellow
coating on the concave side of the
poplar leaf blisters. During summer the
asci release windblown ascospores
which are scattered all over the tree.
These rarely infect existing leaves but
give rise to the yeast-like spores which
enable the fungus to overwinter. Soon
after ascospore release, infected tissues
die and Alnus and Prunus leaves fall
prematurely.

Control: For Alnus and Populus, none
required.
Therapy: None available.
Prevention: (Peach leaf curl) Effective
fungicidal treatments are available for
fruit trees but these are not often likely
to be practicable for ornamentals: spray
with a copper fungicide in autumn,
after leaf fall or with a suitable fungicide (several are available, e.g. captan,
ziram) in late February or early March
(just as buds are swelling). [Take note
of current Pesticides Regulations – see
p.288]
In some cases, such as with wall-trained
fruit trees, it is feasible to prevent the
disease by covering trees from late
January until mid May with a suitably
ventilated polythene shelter, to keep the
foliage dry.

Remarks: Peach leaf curl was first
described in 1821, from England, the
cause demonstrated in 1876, in
Germany; T. tosquinetii was shown to
be the cause of the alder disease in
1923, in Germany.

PEAR AND CHERRY SLUGWORM (PEAR SLUG SAWFLY)   Insect: *Caliroa cerasi*

Damage Type: Removal of leaf epidermis and partial leaf killing.

Symptoms & Diagnosis: From June
onwards pale brown patches appear on
the leaves (Fig.285), this damage
increasing throughout the summer
(Fig.284).
Confirmation: When viewed against
the light, the brown patches are translucent (Fig. 284). A close inspection
reveals that the upper epidermis has
been removed (and occasionally also the
lower epidermis) . Also look for black
and shiny, slug-like, pear-shaped larvae
up to 10mm long (Fig.286); or moulted
skins stuck to the leaf with their front
legs splayed out.

Status: Widespread and common.

Significance: Disfiguring damage to the
foliage of rosaceous trees. Extensive leaf
feeding may lead to premature leaf fall

Fig. 284

Fig. 285

Fig. 286

and a reduction in growth the next year.

Host Trees: Rosaceous trees including Pyrus, Prunus, Crataegus, Sorbus aucuparia and Cydonia. Also reported on Quercus and Salix.

Life Cycle: Several generations per year from June to early autumn. The eggs are laid in slits on the underside of the leaves. The larvae feed exposed on the upper surface of the leaf before dropping down to the soil to pupate in a cocoon. The adult females emerge from these and lay eggs into slits which they cut with their saw-like ovipositor in the underside of the leaves.

Control: Not usually necessary, but if required on small amenity trees use a contact insecticide as soon as the first generation larvae appear in June. [Take note of current Pesticides Regulations – see p.288]

Remarks: Very similar damage in May/June and also in August is caused to rosaceous trees by larvae of Choreutis pariana (the apple leaf skeletonizer) which feed within a silk web on the upper leaf surface (not described further here). See also Caliroa annulipes and C. cinxia on pp.195 and 196.

SECTION 3

PEAR LEAF BLISTER MITE

Mite: *Eriophyes pyri*
Syn: *Phytoptus pyri*

Damage Type: Leaf blistering, discoloration and necrosis.

Symptoms & Diagnosis: As soon as the tree flushes, angular blisters up to about 2mm across appear on the leaves, occurring on both the upper and lower epidermis. Blisters on the lower surface quickly develop a small hole through localised death of the parenchyma cells. The blisters are inconspicuous at first, coloured pinkish to reddish green to yellow (pinkish or reddish on hosts other than Pyrus) but giving the foliage a rough feel. As the season progresses, the blisters darken and contrast with paler brown areas of necrosis (Fig.287).

Status: Widespread and often common.

Significance: Disfigures foliage.

Host Trees: A complex of eriophyid mite species under this name causes similar damage to a wide variety of rosaceous trees including Sorbus torminalis, S. aucuparia (Fig.288), S. aria, Pyrus, Malus and Crataegus.

Life Cycle: The mites overwinter under bud scales and move into the buds as they flush. The galls are initiated *before* the leaves open. Blisters can only be entered by the mites once large enough holes have developed in the galls on the underside of the leaves (see Symptoms above). Blisters that do not develop holes never contain mites. Several generations of eriophyids will occur during spring and early summer, ceasing when the blisters darken and the plant tissue in the galls is dead. The mites then return to the new buds to overwinter.

Fig. 287

Fig. 288

Control: Not necessary.

Further Reading: Jeppson, Kiefer & Baker (1975).

PHOMOPSIS CANKER OF JUNIPER

Fungus: *Phomopsis juniperovora*

Fig. 289

Damage Type: Needle, shoot and bark killing.

Symptoms & Diagnosis: At any time of year, scattered shoots or branches are faded, yellowed, browned or dead (Fig.289).
Confirmation: Cutting into the bark at the base of affected branches reveals the cause of symptoms to be girdling patches of dead bark.
Additional indicators: Resin exudes here and there from affected branches; a few small target cankers, centred on dead side twigs, are present.
Caution: Both Kabatina shoot blight (p.152) and Coryneum canker (p.107) can cause similar symptoms but the latter is uncommon on junipers and the former kills only young shoots.

Status: Widespread and quite common both in planted and wild junipers.

Significance: If neglected this can slowly become very disfiguring on particularly susceptible species and can kill small plants, but infections are sporadic and localised and therefore readily removable. On most species, it merely kills the odd twig or branch from time to time.

Host Trees: Confined to the Cupressaceae (Cypress family) but apart from a few cases on Cupressus macrocarpa, all reports in this country have been from junipers. J. communis is often affected but damage is slight. J. sabina 'Tamariscifolia' seems to be particularly susceptible.

Resistant Trees: Sinclair et al. (1987) state that few if any junipers are immune but list a number of clones of various juniper species which are relatively resistant. J. sabina 'Tam no Blight' is listed in *Hilliers Manual* (6th Edition, 1991) as a resistant form of J. sabina 'Tamariscifolia'.

Infection & Development: The fruit bodies of the fungus are produced on infected needles, shoots and bark. Spores which ooze out of these are carried away in rain water to infect unwounded needles or bark wounds. The fungus soon kills needles, shoots and small twigs and spreads back into older wood but kills this only very slowly. It may therefore produce non-girdling cankers many years old though, in time, it can kill branches an inch or so across.

Control: To prevent the continued spread of an existing infection, in dry weather cut back the infected branch to a side branch or bud below all dead or stained phloem or cambial tissue. Treat the cut with a fungicidal wound paint to guard against reinfection. [Take note of current Pesticides Regulations – see p.288]

Remarks: The disease was first described in 1920, from the USA, and first recorded in Great Britain in 1969.

SECTION **3**

PHYTOPHTHORA BLEEDING CANKER

Fungi: *Phytophthora cactorum and P. citricola*

Damage Type: Bark killing.

Symptoms & Diagnosis: Scattered drops of a rusty-red, yellow-brown or almost black, gummy liquid ooze from small or large patches of bark on stems or limbs (Fig.290). These run a little way down the tree to dry as dark brown or black, often shiny, brittle encrustations or, on the underside of branches, as little pendulous knobbles. (Eventually, limbs may die but the weeping is so conspicuous that it is this that is likely to draw attention to the disease.)

Additional indicators: The centre of an oozing patch of bark may be cracked and perhaps bearing fruit bodies of wood-rotting fungi.

Confirmation: The inner bark of oozing patches is dead or a watery orange-brown colour and often clearly mottled or zoned (Fig.291). The underlying wood may be stained blue-black. **No fungal mycelium is evident under or in** dying bark.

Caution: See also Drought (p.112), Bulgaria canker (p.91), and Coatings, Deposits and Exudates (p.100).

Status: As distinct from Phytophthora root disease (where aerial lesions are the result of fungal growth upwards from roots – see p.206), Bleeding canker seems rather uncommon and has been reported only from the south of the country.

Significance: As the fungus is confined to the bark and is slow spreading it can be eradicated, although later invasion of the wood by decay fungi can still be a problem.

Host Trees: Recorded in this country on Aesculus and Tilia, but there is no apparent reason why other species which are attacked elsewhere (Acer,

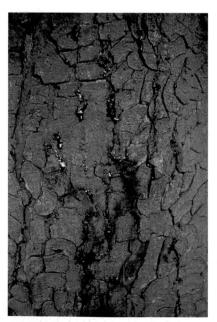

Fig. 290

Fig. 291

Betula, Liquidambar, Quercus, Salix) should not sometimes succumb here too.

Infection & Development: Not yet fully elucidated. As both of these fungi also cause Phytophthora root disease (p.206) they are probably resident in the soil, in roots of other plants on the site or asymptomatically in the roots of Bleeding canker infected trees. In suitably wet and warm conditions, spores would be produced and, by some means not yet determined, reach the aerial parts of the tree, germinate and infect the bark. Infection does not seem to require injury to the bark. The fungus grows through and kills the phloem and cambium and over a number of years may girdle and kill the whole limb or stem. The exuded gum does not contain the fungus.

Control: *Therapy:* Cut out all dead and dying bark. With a fresh or sterilized blade, remove a strip of healthy bark at least 2 inches wide from the periphery of the wound. Treat the exposed tissues with an Approved wound paint based on a fungicide effective against Phytophthora species (e.g. octhilinone or copper). [Take note of current Pesticides Regulations – see p.288] Collect and burn all excised bark. *Prevention:* None available.

Remarks: P. cactorum was, in 1886, the second Phytophthora to be discovered, the first being P. infestans, the cause of Potato blight. P. citricola was described in 1927. Both cause root disease in many crop and ornamental plants. The disease was noted in the USA in about 1930 but not until the 1970s in England.

Further Reading: Strouts, 1981.

SECTION **3**

PHYTOPHTHORA BLIGHT OF HOLLY   Fungus: *Phytophthora ilicis*

Damage Type: Defoliation; leaf blotching; bark killing.

Symptoms & Diagnosis: In autumn or winter, current and older leaves, bearing black patches irregular in size and shape, often only one on a leaf, fall and quickly turn black (Fig.292); defoliation starts at the bottom of the tree and progresses upwards as winter wears on. During winter, black bark patches, often girdling, develop on current and older twigs. In late winter, berries die. In spring and early summer, scattered twigs bearing dying or dead leaves may be evident. At any time of year, sunken cankers may be present on older branches and stems.
Confirmation: This requires laboratory examination during winter.
Additional indicators: Fresh outbreaks follow wet, cool weather. Along the length of a clipped hedge, a number of large, neatly wedge-shaped leafless patches may develop (Fig.293).
Caution: Drought induces the fall of green but unspotted leaves.

Status: Known in gardens and one nursery from a small number of widely scattered localities in the southern half of Great Britain but at the time of writing not reported from woodlands. These few confirmed cases suggest the disease is much more common than is realized.

Significance: Has caused severe damage to isolated hedges and trees, suggesting that it could become a major nuisance.

Host Trees: Ilex aquifolium and its cultivars; in the U.S.A. has also been found on I. colchica, I. pernyi var. veitchii and some clones of I. apaca.

Fig. 292

Fig. 293

Resistant Trees: Of the many species and cultivars tested in the USA, only those mentioned above suffered significant damage.

Infection & Development: The fungus probably survives the dry, warm summer weather as spores (oospores) in leaf and bark tissues it killed the previous winter. In autumn (and possibly in summer if conditions are sufficiently moist and cool), zoospores are produced from the oospores and carried in rain to infect leaves, perhaps through punctures caused by the spines of adjacent leaves. As the fungus grows, black lesions develop, usually first at the leaf margins. Ethylene gas produced during this process causes infected leaves to fall, often when the spots are still small (Fig.292). During winter, more infections arise from zoospores formed on the fresh lesions. Twigs are probably infected through the scars left when diseased leaves fall; they may also be infected when the fungus grows down pedicels from infected fruit. Once girdled, the distal parts of the twigs die (Fig.293). Sometimes the fungus spreads down twigs into stems or larger branches to cause cankers, though these have not been observed in this country. Cankers presumably spread for the first winter only, though this is not clear from the literature. If infections are confined to the leaves, defoliated twigs will refoliate in the following year. Long distance spread will be effected by the transport of infected planting stock, but the development of the disease in this country since the first case came to light suggests some other means is involved.

Control: *Therapy:* None available.
Prevention: The application of a Potato blight (Phytophthora infestans) fungicide at fortnightly intervals during autumn should be effective. [Take note of current Pesticides Regulations – see p.288]

Remarks: The disease was present in the USA in the 1930s but the cause only determined in the early 1950s. It was confirmed in Holland in 1988 and in Great Britain, in Sussex, in 1989 on a tree which had shown symptoms of the disease for some years previously.

PHYTOPHTHORA DISEASE OF ALDER

Fungus: *Phytophthora x cambivora*

Fig. 295

Fig. 294

SECTION 3

Damage type: Bark killing on stems and roots

Symptoms and Diagnosis: Foliage is yellowish, sparse and small (Fig. 295). Coning is often heavy. Lower part of stem shows tarry or rusty exudates (Fig. 294). Trees can die rapidly over a single season or deteriorate slowly over a number of seasons. In some instances epicormic shoots on the main stem remain alive even after all the branches have died.

Additional indicators: The site is subject to flooding from an adjacent river.

Confirmation: The inner bark below the tarry patches is discoloured, often mottled or zoned. Most commonly columns of dead bark extend upwards from ground level but very occasionally lesions are of the bleeding canker type (q.v. p.202).

Caution: Tarry spots are quite often found on the stems of alders suffering from severe drought damage. Excavation of the bark will reveal only localised necrosis.

A crown dieback of alder of unknown cause is common in riverside trees in Scotland and northern Britain (See Comment 1 under Alnus, p.271).

Status: Common and serious through much of Britain. Only a few records have been made in Scotland.

Significance: Over the period 1994–1999, mortality of alder on riverbanks in southern Britain increased at between 1 and 2% per annum.

Host trees: Most cases concern the common alder Alnus glutinosa. However, A. cordata, A. incana and A. rubra are all susceptible.

Infection & Development: The disease seems to be principally disseminated by water, with self-propelled spores probably playing an important part. The precise method of infection is not known. Once within the tissues the fungus can grow at a rate of about 1 cm per week. Trees can recover, possibly when high summer temperatures are inimical to the fungus.

Control: None available at present.

Remarks: The pathogen is now known to be a hybrid between two Phytophthora species neither of which can attack alder. These are Phytophthora cambivora, a common pathogen of broadleaved trees and a species closely resembling P. fragariae, the cause of Red core disease in the roots of strawberry. Although not known prior to 1993 it must have been active in the U.K. for at least 30 years before that date. It is known now also to be widespread in Europe, although the amount of damage varies greatly from place to place.

Further reading: Gibbs and Lonsdale, 1998.

PHYTOPHTHORA ROOT DISEASES

Fungi: *Phytophthora species; mainly P. cambivora, P. cinnamomi, P. citricola*

Damage Type: Killing of roots and stem base.

Symptoms & Diagnosis: (Figs 296-301) A tree dies suddenly or slowly (Fig.296), or shows symptoms indicative of a root disorder (see Table III on p.14 and Fig.298 – compare with healthy tree in background); or all (or only the lower) branches on one side of a tree die (Fig.297).
Confirmation: Characteristic, though not always present, are tongues of dead bark extending a few inches to several feet up the stem from dead roots (accounting for the one-sided branch killing – Figs.299-301). Roots, and probably the stem base, are dead but sound.

Fig. 296

Fig. 297

Additional indicators: Fungal fruit bodies, mycelium or other fungal structures are absent; resin or gum oozes from the lower stem; the site is liable to lie wet, is irrigated heavily (Fig.296), or is heavily mulched with farmyard manure or compost; other susceptible species on the site are also affected.

Caution: Easily confused with waterlogging but dead, waterlogged roots are blue-black and, like the soil, malodorous. See also bonfires, lightning and de-icing salt damage (pp.122, 170 and 93) and Remarks below. The killed bark is sometimes quickly invaded by Honey fungus.

Status: Common and widespread in the south but becomes less common northwards.

Significance: A frequent cause of debility, dieback and death of amenity trees but often overlooked or misdiagnosed.

Host Trees: Recorded on a very wide range. Notably susceptible are Castanea (on which the disease is known as Ink disease – Figs.298,299,300), Chamaecyparis lawsoniana (Figs.297,301), Eucalyptus, Nothofagus; also Taxus, on which it is the only known fatal disease in this country Fig.296). Also a prominent disease of Aesculus, Azalea and Rhododendron, Calluna and Erica, Fagus, Malus in orchards (we have rarely seen the disease in crabapples), Tilia and Prunus (Flowering cherries). Severe damage to Abies species has been noted in nurseries and in plantations a few years old but reported cases on older trees are rare.

Resistant Trees: Any commonly planted species not mentioned above may be regarded as relatively resistant. Rarely recorded on × Cupressocyparis leylandii, Quercus petraea or Q. robur.

Infection & Development: Spores, moving in soil water partly passively, partly self-propelled, infect roots and basal stem bark. Mycelium then grows through and kills the cambium and phloem. Trees die through the loss of functioning roots or when the stem is girdled at the root collar. The fungus persists in the soil and is thus readily transported on plant roots, machinery, tools and footwear. It is also harboured and carried in water (streams, irrigation ponds, tanks). Trees can recover but remain open to further attack. Several species of Phytophthora cause identical diseases. Most have host preferences though some have a very wide host range. Their temperature requirements differ though all require free water for spore production, spread and germination so the disease is commonest on

SECTION **3**

Fig. 298

Fig. 299

Fig. 300

Fig. 301

sites which are liable to lie wet. It is also encouraged by the presence of large quantities of decaying organic matter such as lawn mowings, farmyard manure and compost.

Control: *Therapy:* None available.
Prevention: (i) The disease is often brought in on planting stock, so use plants from Phytophthora-free nurseries. (ii) Irrigate in moderation; do not allow water to pool around susceptible species. (iii) Replant with resistant species. Fungicides based on fosetyl aluminium and metalaxyl are useful in combatting the disease in orchards and could conceivably have a use in future in amenity plantings. [Take note of current Pesticides Regulations – see p.288]

Remarks: These diseases are among the world's most destructive, affecting an enormous range of crops and natural ecosystems. The first cases in this country to be confirmed were on woodland Castanea and Fagus in the 1930's but the diseases have been recognised as important on amenity trees here only since about 1970. Root and basal stem-bark killing in trees damaged by deicing salt may in fact be due to Phytophthora – see Dobson (1991d).

Further Reading: Strouts 1981.

PINE SHOOT BEETLE

Insect: *Tomicus piniperda*
(= Myelophilus piniperda)

Damage Type: Localized destruction of main stem cambium.

Symptoms & Diagnosis: Trees die.
Confirmation: From June onwards a scatter of insect emergence holes about 1.5mm diameter ('shot holes') appears in the thicker bark of these dead or dying trees; on live trees, from March onwards, beetle entry holes are often indicated by resin tubes on the trunk (Fig.303). If pieces of the perforated bark are levered off, bark beetle galleries are revealed in the bark cambium (Fig.304).
Additional indicators: The current shoots on nearby trees become brown late in the year (or are sometimes snapped off earlier, and while still green, by strong winds) (Fig.302). A closer examination will reveal a tunnel through the centre of such shoots with an entrance hole that may have a resin tube surrounding it (Fig.305). These tunnels are caused by adult beetles feeding in summer and autumn (Fig.306); such behaviour is known as shoot pruning.
Caution: Resin tubes at the base of live trees in the winter show that beetles are present but do not necessarily indicate a health problem (see Life Cycle below).

Status: The insect is common and widespread.

Significance: The majority of attacks by T. piniperda are secondary to another problem. This beetle usually only breeds in standing trees if they have been weakened or killed recently by another agent such as lightning, fire, windblow, drought or disease. T. piniperda is rarely, if ever, involved in

SECTION **3**

Fig. 302

the death of trees that are in full health and vigour. Semi-mature ball-rooted trees in amenity plantings that are stressed due to water deficit can be attacked and killed if an endemic population of the beetle exists nearby. Shoot pruning in summer does not predispose trees to beetle attack for breeding purposes although it may result in poor tree form.

Host Trees: Pinus. Less resinous species such as P. sylvestris are more susceptible.

Life Cycle: Adult beetles (dark brown or blackish and 3–5mm long) bore longitudinal galleries in the bark cambium of felled, windblown or severely stressed trees from March to May (Fig.304). These mother galleries can score the sapwood on thin barked stems of small diameter. The female beetles lay eggs in niches along either side of the mother galleries from which larval tunnels subsequently radiate out, widening as the larvae develop, and ending in an oval cell where they pupate. Development takes 12–13 weeks from egg to adult and the new generation of T. piniperda emerge from mid-June to August, with the peak in mid-July. It is mainly these freshly emerged beetles that tunnel into current shoots and cause 'shoot pruning' (see above). The adult beetles overwin-ter either in the hollow shoots (fallen or still on the tree), or in short tunnels which are bored into thick bark at the base of healthy pines, often giving rise to white resin tubes.

Control: *Therapy:* Debark or remove from the vicinity freshly cut logs as these are potential breeding material. Most breeding is in cut logs less than six months old, or in blown trees which still have some intact roots and green foliage.

Prevention: No chemical is recommended for treating live trees. Chlorpyrifos is approved for spraying infested logs but such treatment can rarely be justified and is not recommended. [Take note of current Pesticides Regulations – see p.288]

Further Reading: Winter (1991), Bevan (1987), Carter & Winter (1998).

Fig. 303

Fig. 304

Fig. 305

Fig. 306

PINE SHOOT MOTH   Insect: *Rhyacionia buoliana* (= *Evetria buoliana*)

Damage Type: Destruction of interior of shoots and buds.

Symptoms & Diagnosis: The leading shoots are deformed (Fig.307) or killed and the terminal and/or lateral buds fail to flush.

Confirmation: In spring and autumn newly constructed tents of silk and resin between infested buds (Fig.308), or at the base of mined shoots, glisten in bright sunshine and advertise the presence of the dark reddish or purple-brown larvae.

Additional indicator: Moths are silver with orange or ferruginous markings, wing span 16–24mm. They are nocturnal but are often disturbed from the canopy during the day when they fly to a nearby tree or flutter to the ground.

Status: Widespread throughout England reaching the Scottish border in the west. Also in Wales and Dumfriesshire.

Significance: Most important in young forest plantations. In amenity plantings and gardens this insect can cause moderately severe damage, especially to dwarf cultivars. Continuous heavy attacks will cause multiple shoot development resulting in 'stag-headed' leaders and stunted trees. Damage to recently flushed leading shoots can cause these to bend over and if they then recover dominance, the tree may be left with a 'posthorn' shaped deformity in the main stem.

Host Trees: Most Pinus spp. In forestry the fast growing species such as P. muricata and P. radiata are particularly susceptible, as also is P. contorta. Resinous species such as P. nigra maritima are resistant. Reported once on Picea breweriana.

Fig. 307

Fig. 308

Life Cycle: Adult moths occur in late June and July when the eggs are laid on needle sheaths close to the apical buds. When a larva first hatches it mines the base of a needle pair and then moves to a lateral bud. Here a small shiny tent is constructed from silk and resin after which the bud is mined (more than one may be attacked if the buds are very small). The larva overwinters in a mined bud with the entrance hole sealed with silk. In March or early April the larva moves to a leading bud or shoot where a new resin tent is constructed. The larva then feeds in the bud, or tunnels the shoot if the tree has flushed. It pupates in the last feeding place.

Control: Rarely practicable. Two applications of a potent permitted contact insecticide, applied two weeks apart will give good control if timed accurately to coincide with hatching of the eggs in August. [Take note of current Pesticides Regulations – see p.288]

Further Reading: Scott (1972), Carter & Winter (1998).

POCKET PLUMS   Fungus: *Taphrina pruni*

Damage Type: Deformation and destruction of fruit; deformation of fruiting branches.

Symptoms & Diagnosis: The fruits (plums, cherries etc.) are peculiarly swollen, elongated and perhaps curved, like tiny bananas, up to about 6 cm long and 2 cm thick (Fig.309). They remain pale yellowish, sometimes tinged pink. instead of ripening and during summer become wrinkled and covered in a whitish or greyish-pink bloom (asci full of spores). In late summer and autumn they shrivel and some fall but others remain on the tree during winter. Often, all the fruit on certain branches is affected while the rest of the tree is normal. Twigs bearing the deformed fruit also tend to be thickened and deformed.
Confirmation: In most of the deformed fruits there is no stone, only a cavity.

Status: Occasional, commonest in the English West Country.

Significance: On ornamentals, an interesting curiosity; in fruit growing, an occasional, readily controlled problem.

Host Trees: The plums and similar fruit: Prunus domestica and subspecies insititia (damson), P. cerasifera, P. padus, P. spinosa and a few other, rarely grown species.

Infection & Development: Not entirely clear from the literature, but it seems that, like other Taphrina species (cf. Taphrina witches' brooms, p.270), the ascospores shed in summer from the diseased fruit give rise to non-infective spores which survive winter on bark and among bud scales. In spring,

Fig. 309

these infect emerging shoots (perhaps via leaves) and a perennating mycelium becomes established in the bark. In subsequent springs, this invades flower buds and the developing ovary to produce the 'pocket plums' on the infected branches year after year.

Control: None required for ornamentals but the thorough removal of infected branches and fruit will quickly eliminate the disease.

Remarks: The disease was described by an Italian botanist in 1583 [sic].

SECTION   3

POPLAR CLOAKED BELL MOTH

Insect: *Gypsonoma aceriana*

Damage Type: Hollowing out (mining) and consequent death of current shoots.

Symptoms & Diagnosis: Death, loss of vigour or premature cessation of growth in scattered shoots in May or June.
Confirmation: An erect conical tube of frass (larval excreta and other debris) projects from a shoot/petiole junction on the affected shoot. If split longitudinally, the shoot is seen to be mined upwards from this point (Fig.310).

Status: Widespread and locally common in southern England and south Wales, scarce in northern England and north Wales, rare in Scotland.

Significance: Little more than a curiosity on amenity trees but sometimes causes significant damage in stool beds and nurseries where failure of the leading shoot results in subsequent forking in young crops.

Host Trees: Only on Populus species including P. nigra, P. nigra 'Italica', P. alba and P. balsamifera. There is some evidence that the earlier flushing clones such as 'Balsam Spire' are the most susceptible.

Life Cycle: This has not been fully worked out for Britain. In May and June the larvae mine the petioles and then the shoots; they pupate in July either in bark crevices or in the litter. Adult moths emerge and fly in July and August.

Control: Not necessary on amenity trees.

Remarks: On the Continent, G. aceriana sometimes causes stem galls or mines buds but the Forestry Commission has not received reports of such damage from Britain.

Fig. 310

POWDERY MILDEWS	Fungi: *Mainly Microsphaera, Podosphaera, Sawadaea (= Uncinula) and Phyllactinia species*

Fig. 311

Fig. 312

Damage Type: Killing of soft tissues; sometimes defoliation.

Symptoms & Diagnosis: Leaves, soft shoot tips and perhaps young flowers and fruitlets have a complete or patchy white (or pale brown), thin, mould or powder-like coating, perhaps associated with necrotic patches and premature defoliation (Figs.311-313); affected parts may be distorted and dwarfed. *Confirmation:* Under a hand lens, the white patches are seen to consist of tangled threads (hyphae) strewn with powder (spores).

Status: Widespread and common.

Significance: On established ornamentals, this is mainly an aesthetic problem but small nursery trees can be stunted or killed and fruit production in fruit trees is reduced.

Host Trees: Many broadleaved species. Perhaps the commonest Powdery mildews are on Quercus (M. alphitoides – Fig.312), Malus (P. leucotricha – Fig.311), Crataegus (P. clandestina var. clandestina), Acer (S. bicornis) and Corylus (Ph. guttata). The disease on Platanus (M. platani), has been occasionally conspicuous in places in England since its discovery here in 1983 and may become common. There are rare reports of severe damage to Prunus laurocerasus from the Rose mildew fungus, Sphaerotheca pannosa – see Remarks below.

Resistant Trees: Conifers are immune. See also Control below.

Infection & Development: Most species overwinter as mycelium in the buds, from which shoots and leaves emerge in spring already infected. The species on Acer and Corylus overwinter as fruit bodies fallen from infected leaves and adhering to buds and twigs. In spring, ascospores from these infect the emerging leaves and shoots. The infected tissues quickly produce airborne spores (conidia) which (unusual among fungi) do not require moisture for germination. These infect more new tissues and this cycle continues all the while new, soft tissues are available. At first, the mainly superficial mycelium withdraws water and nutrients from epidermal plant cells without killing them but eventually necrosis, distortion and defoliation results.

Control: *Therapy:* The Powdery

SECTION **3**

mildews are among the few diseases which, in their early stages, can be cured with fungicidal sprays, as their mycelium is mostly superficial (except in Ph. guttata) though such treatment is justifiable only on nursery and fruit trees.

Prevention: Avoidance of pollarding, hedge trimming etc. will reduce the incidence of these diseases by limiting the amount of susceptible, soft tissues available during the summer period when conidia are most abundant. Effective fungicides are available for nursery and fruit trees but their use is not warranted on amenity trees. [Take note of current Pesticides Regulations – see p.288]

Remarks: Several species of Powdery mildew occur on but cause little damage to Prunus species but severe damage has been noted on P. laurocerasus from the Rose mildew, Sphaerotheca pannosa (Fig.313). Current leaves crinkle, develop brown patches which disintegrate to leave ragged holes, and fall. Some leaves fold upwards along the midrib, the pale undersides giving a false impression of the Silver leaf disease (p.252).

The pathogenicity of Powdery mildews was first demonstrated in the 1860s, in Germany. Oak mildew was first recorded in Europe in 1907 (perhaps 1877), in

Fig. 313

England in 1908; Plane mildew was recorded in Europe in the 1970s but the first report from Britain was from London in 1983. Both species originated in the USA. The other species were known in Europe in the 19th century.

Fig.311 also shows the Green apple aphid, Aphis pomi.

| PRUNUS KANZAN DIEBACK | Cause: *Uncertain; possibly the bacterium Pseudomonas syringae* |

Damage Type: Bark killing.

Symptoms & Diagnosis: In early summer, the foliage on one or several branches wilts suddenly and slowly turns brown but does not fall (Fig.314). The symptoms may spread in the same year or in the next to the whole crown.

Confirmation: Depending on the age of the damage, either all or most of the bark of the wilted branch is dead or the outer phloem is alive while the inner phloem and cambium is stained brown. (If most bark is alive, the branch base should be checked – it may be girdled as the result of dead bark extending from adjacent branches.) Dead bark may extend down onto the stem. The wood underlying the dead bark is not stained. The rootstock may survive and sprout.

Fig. 314

Additional indicators: The symptoms follow particularly cold winter weather. Other affected trees occur in the locality. *Caution:* Can be confused with Bacterial canker, but gummosis is absent (p.70); or with lightning damage (p.170).

Status: Widespread following cold winters in England and Wales. Not reported from Scotland.

Significance: A severely disfiguring and often fatal disease.

Host Trees: All cases investigated have been on Prunus 'Kanzan' or on similar but unidentified pink, double flowered Flowering cherry cultivars.

Infection & Development: The symptoms indicate that a bark killing pathogen spreads in the inner phloem and cambium from an indeterminate entry point in the crown, thence down branches, often into the main stem. No pathogen has been isolated or identified but symptoms suggest a bacterial cause. Its association with very cold winters suggests Pseudomonas syringae pv. syringae to be a likely candidate as infection by and spread of this known pathogen of Prunus is encouraged by freezing temperatures.

Control: *Therapy:* The removal of affected branches by cutting well below any dead bark or stained phloem should prevent further spread, though often symptoms are noticed too late, after the main stem is involved.
Prevention: In planting schemes, too much reliance should not be placed on Prunus 'Kanzan'.

Remarks: The best hope of identifying the cause is for symptoms to be reported to pathologists very early in the year, when the pathogen may still be active and therefore detectable. Why the disease should have come into prominance only after the 1981/2 winter is unknown; it seems unlikely that earlier cases would have gone unreported.

Further Reading: Strouts, 1991b.

SECTION **3**

ROTS OF ROOTS AND BUTT
(See also Section 5)
Fungi: *Various species, mostly bracket and gill fungi*

Damage Type: Wood decay with or without bark killing.

Symptoms & Diagnosis:

• The crown of the tree is thin (i.e. less dense than to be expected for the species); leaves are generally small and pale (Fig.315 – cf healthy neighbouring tree of same species); a scattering of dead twigs and branches or branch ends may be present; or

• The tree has flushed much later and more feebly than usual, or in autumn has shed its leaves much earlier than usual; or

• The tree is dead or almost dead or has fallen over; or

• Fruit bodies of a known root- and butt-rotting species are growing from the base of the trunk, from buttress roots or from the ground over the roots (Figs.183,316-335,345,346).

Confirmation: Parts of the stem base below (and perhaps also above) ground level and/or some major roots are found to be dead and either decayed or the bark permeated by or overlying fungal mycelium.

Additional indicators: The growth rate of the tree (shoot extension, annual ring widths) has fallen off markedly in the last few years; adjacent trees also show or have shown such symptoms.

Caution: Similar crown symptoms result from the death or malfunction of roots from whatever cause e.g. Phytophthora root disease (p.206), flooding (p.266), poisoning (p.93); or from the girdling or strangulation of the stem below the live crown as a result of damage from e.g. machinery or animals (p.176); or from site alterations (p.177). Some harmless fungi produce fruit bodies resembling those of damaging species.

Fig. 315 *Sorbus aria* dying from *Heterobasidion annosum* root infection

Status: This type of disease is common and widespread, though some of the causal fungi are not.

Significance: Among the most damaging of diseases, particularly in older trees: crown deterioration spoils the tree's appearance and screening capability; decay may render the tree unsafe; ultimately the tree may blow over or die. The identification and the significance of the various root- and butt-rotting fungi is dealt with below.

Host Trees: It is doubtful if any species is immune from attack by every one of our many root- and butt-rotting fungi but some are particularly prone to such disease, Fagus sylvatica among them. See also Remarks below and Section 2.

Infection & Development: Of the many root- and butt-rotting fungi in this country, the infection biologies of only the Armillaria species (Honey fungus, p.145, and Fig.316), Phaeolus schweinitzii (Greig 1981) (Fig.330) and Hetero-basidion annosum (see below and Greig 1981) have been fully worked out.

It is probable that most of the remainder infect tree roots by means of airborne spores washed into the soil. The spores are produced in fruit bodies which arise either directly from the stem base or from exposed roots (e.g. the inconspicuous Ustulina deusta – Fig.333), or grow out of the ground over infected roots, attached to them by mycelial strands (e.g. Meripilus giganteus – Fig.329). It is also likely that some species can spread to neighbouring trees by growing out of an infected root into a contiguous root of the healthy susceptible tree.

After infecting the root, the fungus grows through it, both killing and later decaying it. Depending on species, this killing may extend above ground into the basal stem bark, but often the fungus above ground is confined to the heart wood. Some species affect few roots but cause extensive decay in the butt, and some extend far up the tree to cause stem rots. The rate of spread and therefore of root death and decay will depend on the species of fungus and tree, the tree's vigour and environmental conditions. The tree may survive for many years without obvious distress but eventually crown symptoms begin to appear. This happens when root killing has outpaced root regeneration and the reduced root system can no longer supply the demands of the fully foliated crown. This point may be reached gradually or suddenly, for example in a period of drought, and the progression of symptoms may be correspondingly slow or fast.

Sometimes, as a result of the root killing and decay combined with such factors as wind, tree crown weight and soil type, the root anchorage fails and the tree falls over; or the decay in the butt may be so extensive that the stem snaps. After the death or fall of the tree, the fungus is likely to persist saprophytically in the stump for some years, dying only when all the nutrients are exhausted and the wood thus destroyed.

Some of the root- and butt-rotting fungi also infect above-ground wounds to cause branch and top rots. Occasionally such infections spread downwards into the stump and roots.

Control: *Where tree safety is the prime consideration*: Tree hazard assessment, the detection of decay and available control measures are discussed briefly in Section 5 in this book and in detail in Lonsdale (1999). *To prevent further spread of the disease*: For Armillaria (Honey fungus) see p.148. For Heterobasidion annosum, where conifers are felled within root contact of existing susceptible species and the stumps cannot be removed, paint the freshly cut stump surface liberally with creosote or, if to hand, the material currently approved for use for this purpose in forestry, otherwise either grub out infected trees or stumps before replanting or avoid planting susceptible species within root contact of them. This latter simple precaution could sensibly be taken also in the case of the rest of these root diseases, even though their infection biology is not or is only imperfectly understood.

Remarks: See also Section 5, Decay and Safety.

The little information available on many of these diseases is derived largely from *ad hoc* observations rather than from deliberate research. Consequently their infection biology is often incompletely known or quite unknown. Similarly, the effect on the tree of some fungi which look as if they might be root or butt rotters (because they fruit at the base of trees) is very uncertain.

SECTION 3

The effect on the tree of the various root- and butt-rotting fungi varies hugely depending on the species of fungus and tree involved. The identity of the fungus can therefore often be invaluable in prognosis and in deciding on appropriate control measures. Identification from the decay alone usually requires laboratory techniques but if fruit bodies are associated with the symptomatic tree, the fungus can often be identified from these. A means for identifying in the field all of the species likely to be encountered and notes on their significance are presented on p.226.

On many species:

Armillaria (Honey fungus): see p.145.
Heterobasidion annosum ('fomes') (Fig.325) has been studied extensively in its capacity as a major disease of commercial coniferous forests. It is also an occasional problem of broadleaved and coniferous amenity trees (Fig.315, killing a Sorbus aria), spreading via root contacts from infected trees or stumps into healthy trees. New centres of infection arise when airborne spores alight on and infect freshly cut stumps (usually conifers, occasionally Betula). The fungus can infect a huge range of trees and shrubs; genera mentioned in this book on which it has not been recorded and which would therefore be worth considering for infected sites are indicated in the Host-Agent Directory. For control, see above.
Coniophora puteana (= C. cerebella). (Fig.319) Although there is some evidence that this fungus can cause a root- and butt-rot, it is usually saprophytic. It occurs on many species of standing tree but seems to have some preference for Cedrus, Cupressus and Taxodium. It inhabits stumps and woody debris, and decays wood of standing trees exposed by mechanical injuries or via bark killed by, for example, fire, lightning or disease. Its thin, amorphous, often extensive fruit bodies may develop

harmlessly on the boles of healthy trees, sometimes spreading from the litter.

On Fagus:

Meripilus giganteus, the Giant polypore (Fig.329), is a common cause of the death and/or windblow of mature beech trees. The appearance (in late summer or autumn) of its fruit bodies at the base of the trunk or over the root spread should be taken as a sign that the tree is probably or soon will become unstable. The more numerous the clumps of fruit bodies the greater the risk, while if the crown is also thin or dying back the tree should be regarded as an immediate danger, even if no rotten roots can be found: the fungus destroys the deeper-going roots, commonly leaving the shallower, accessible roots intact and healthy to the very last. Its infection biology is unknown.

On Quercus and Castanea:

Collybia fusipes (Fig.317) kills the roots and the below-ground portion of the stem base and rots the sapwood of various oaks and probably occasionally of Sweet chestnut. This results in crown dieback, but reported cases of windblow involving this fungus are rare.

Various broadleaves:

Pholiota species. The commonest British species, P. squarrosa (Fig.331) is occasionally the apparent cause of pruning wound decay in this country and causes a butt-rot in poplar in the USA but its significance when found at the base of trees here is unclear: on ash its toadstools have been found growing from apparently virtually healthy roots, the only dead tissue being a very shallow and discrete patch of outer bark. The presence of this or other Pholiota species should therefore encourage a close inspection of a tree but not be taken to indicate certain decay.

Further Reading:
Lonsdale (1999); Mattheck & Breloer (1994).

Fig. 319 *Coniophora puteana*

Fig. 316 *Armillaria* species
(Honey fungus)

Fig. 317 *Collybia fusipes*

Fig. 318 *Fistulina hepatica* (Beef steak fungus)

SECTION 3

Fig. 320 *Ganoderma adspersum*
(G. applanatum is similar*)*

Fig. 322 *Ganoderma pfeifferi*

Fig. 321 *Ganoderma resinaceum*
(G. valesciacum is similar*)*

Fig. 323 *Grifola frondosa* (with a
Fistulina hepatica bracket above it)

Fig. 324 *Hypholoma fasciculare* (Sulphur tuft)

Fig. 325 *Heterobasidion annosum* (Fomes)

Fig. 327 *Inonotus dryadeus*

Fig. 326 *Leptoporus ellipsosporus on Cupressus (See 'Comment' on p.34)*

SECTION 3

Fig. 328 *Perenniporia fraxinea (= Fomitopsis cytisina)*

Fig. 329 *Meripilus giganteus* (Giant polypore)

Fig. 330 *Phaeolus schweinitzii* (note old, blackened fruit bodies in the background)

Fig. 331 *Pholiota squarrosa*

Fig. 332 *Rigidoporus ulmarius*
(broken open to show dark tubes but pale flesh)

Fig. 333 *Ustulina deusta* – teleomorph

Fig. 334 *Sparassis crispa*
(Cauliflower fungus)

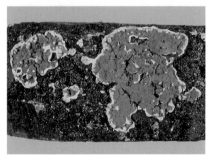

Fig. 335 *Ustulina deusta* – anamorph

SECTION 3

THE IDENTIFICATION OF ROOT, BUTT AND TOP-ROTTING FUNGAL FRUIT BODIES

NB: *This guide works only for the genera mentioned in this book. There are many other species which occur on standing trees – mostly rare or of little importance to tree health or safety – which can be identified only with the aid of a much more comprehensive guide.*

Procedure for using the guide

Step 1:   Note fruit body type (e.g.Bracket with pores – see Chart on p.227).

Step 2:   Note location of fruit body (root- and butt-rotters fruit over roots, at tree base, on bole or, occasionally, higher up the trunk; top-rotters fruit on trunk or branches, often on exposed wood, on cankers or on dead bark).

Step 3:   Refer to Chart and draw up list (1) of possible species according to fruit body type and location (root- and butt-rotters are printed in italics; top rotters are printed in bold).

Step 4:   Identify tree and find its genus in Section 2 of this book.

Step 5:   Draw up list (2) of the root- and butt- or top-rotting species given there.

Step 6:   Compare List (1) with List (2) and note any species which appears in both.

Step 7:   Confirm identity by comparing your specimen with the description of this fungus/these fungi in Table v and with the illustrations.

The Identification of the Principal Wood-Rotting Fungi on Standing Trees

1. Species in italics are root- and/or butt-rotters
2. Species in **bold italics** are top-rotters (note that some can be either)
3. The numbers correspond with the numbers in Table v, column 1.

N.B. This chart works only for the species given in Table v

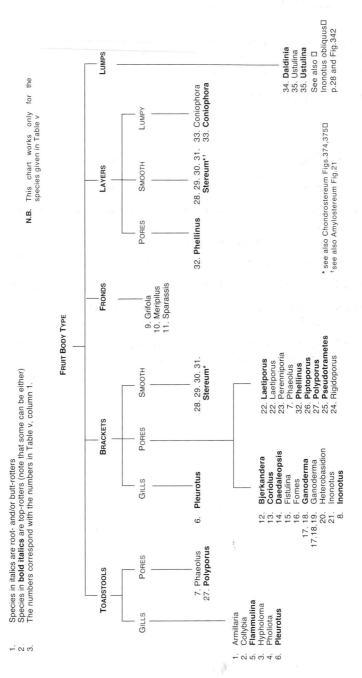

FRUIT BODY TYPE

TOADSTOOLS

GILLS

PORES
7. Phaeolus
27. **Polyporus**

1. Armillaria
2. Collybia
5. **Flammulina**
3. Hypholoma
4. Pholiota
6. **Pleurotus**

BRACKETS

GILLS

6. **Pleurotus**

12. **Bjerkandera**
13. **Coriolus**
14. **Daedaleopsis**
15. Fistulina
16. Fomes
17. 18. 19. **Ganoderma**
17. 18. 19. Ganoderma
20. Heterobasidion
21. Inonotus
8. **Inonotus**

PORES

22. **Laetiporus**
22. Laetiporus
23. Perenniporia
7. Phaeolus
32. **Phellinus**
26. **Piptoporus**
27. **Polyporus**
25. **Pseudotrametes**
24. Rigidoporus

SMOOTH
28. 29. 30. 31.
Stereum*

FRONDS

9. Grifola
10. Meripilus
11. Sparassis

LAYERS

PORES
32. **Phellinus**

SMOOTH
28. 29. 30. 31.
Stereum*†

LUMPY
33. Coniophora
33. **Coniophora**

LUMPS

34. **Daldinia**
35. Ustulina
35. **Ustulina**

See also □
Inonotus obliquus□
p.28 and Fig.342

* see also Chondrostereum Figs.374,375□
† see also Amylostereum Fig.21

SECTION
3

The Identification of the Principal Wood-Rotting Fungi on Standing Trees

Table v.

1. The species are arranged by type (toadstools; brackets; fronds; layers; lumps) and in alphabetical order within these groups. EXCEPT THAT some similar-looking species have been grouped together so that the differences between them are evident.

2. (R) = Root and/or butt rotter. (L) = Likely to render tree unsafe. (M) = May render tree unsafe. (N) = Not known to render trees unsafe. (D) = More likely to cause dieback or death than dangerous decay. (T) = Top rotter. (S) = May cause stem or branch snap. (C) = Rot usually confined to small pockets or dead branches

Fungus	Top/Underside	Texture	Flesh	Tubes/Gills	Other Distinctive Features	Dead State	Decay Characteristics
1. Armillaria species Figs. 183, 316 (R)(M)(D)	Top honey yellow to brownish (Sometimes scaly).	Soft, not tough.	White. Astringent taste.	Gills whitish, later brownish.	Ring on stem below cap. No volva. Spores white.	Quickly rots away	Wet. Brown or yellowish with black lines.
2. Collybia fusipes Fig. 317 (R)(N)(D)	Top brownish.	Soft but tough.	Whitish. Mild taste.	Gills whitish.	Stalk brown grooved, tapering downwards, tough.	Slow to rot away.	Dry, yellowish, confined to sapwood of roots.
3. Hypholoma fasciculare Fig. 324 (R)(N)	Top yellow (often with reddish tinge in centre)	Soft, rather fragile.	Bitter taste.	Gills yellow with greenish tinge, later purplish brown.		Quickly rots away.	Dry, fibrous yellowish.
4. Pholiota squarrosa Fig. 331 (R?) Significance uncertain	Top yellow, very scaly.	Soft not tough.	Bitter taste.	Gills yellowish, becoming brown.	Ring on stem below cap; stem very scaly. Spores brown. No volva.	Quickly rots away.	(Apparently not always associated with decay).
5. Flammulina velutipes Fig. 344 (C)	Top yellowish, slimy.	Soft.	Yellowish. Pleasant taste.	Gills yellowish.	Velvety dark brown stalks. Withstands freezing temperatures.	Slow to rot away.	No information.
6. Pleurotus ostreatus Fig. 349 (C?)(S)(T)	Top shiny brown to blue-grey or violet.	Soft	White. Mild taste.	Gills white to yellowish.	Very short, lateral stem.	Quickly rots away.	White, flaky, surrounded by dark brown narrow band.
Polyporus squamosus See No. 27							

Toadstools

Fungus	Top/Underside	Texture	Flesh	Tubes/Gills	Other Distinctive Features	Dead State	Decay Characteristics
7. Phaeolus schweinitzii Fig. 330	Top dark brown, yellow rim, velvety. Pore surface yellowish green.	Soft, wet, spongy, later dry and hard.	Rusty brown.	Tubes yellowish to rusty brown.	May be found as brackets or on ground with stalk, like toadstools. Confined to conifers.	Dry, dark brown, "resembling old cow-pats".	Crumbly, brown, cubical (often with yellow mycelium in cracks). Smells of turpentine.
8. Inonotus hispidus Fig. 341, 343	Top rusty brown, bristly. Pore surface rusty brown.	Soft, wet, spongy, later dry and hard.	Rusty brown.	Tubes rusty brown.	Never stalked, never growing from the ground.	Dry, black, cracked.	Spongy, yellow-brown surrounded by gummy brown zone.
9. Grifola frondosa Fig. 323	Top grey or greyish brown. Pore surface whitish.	Soft.	White. Pleasant taste, later acrid. Said to smell of mice.	Tubes.	Does not bruise brown like Meripilus giganteus. Fronds up to 6 cm, clumps up to 30 cm across. Uncommon.	Quickly rots away.	White, stringy, with orange lines.
10. Meripilus giganteus Fig. 329	Top yellowish brown with darker brown concentric zones. Pore surface cream.	Soft.	White. On drying, strong, unpleasant odour develops. Sour taste.	Tubes.	Pore surface bruises blackish when handled. Fronds up to 20 cm, clumps up to 60 cm across. Common	Quickly rots away.	Dry, fibrous whitish, with small white pockets.
11. Sparassis crispa Fig. 334	Fronds cream-coloured, discolouring brown with age.	Soft.	White.	None.	Whole fruitbody is reminiscent of a large sponge or open cauliflower. Confined to conifers.	Quickly rots away.	Crumbly, brown cubical (often with white mycelium in cracks). Much like Ph. schweinitzii.

Toadstools ◄——— ———► Fronds

SECTION 3

Brackets

Fungus	Top/Underside	Texture	Flesh	Tubes/Gills	Other Distinctive Features	Dead State	Decay Characteristics
12. Bjerkandera adusta Fig. 338. (T) (C) but (M) in Fagus.	Top smooth, whitish. Pore surface dark grey.	Tough.	Sour taste.	Tubes.	Forms tiers of small thin brackets. Very common.	Slow to decay. Top and pore surface black.	Whitish, dry.
13. Coriolus versicolor Fig. 336. (C) (T)	Top velvety, concentrically zoned, greyish, brownish, reddish or blackish. Pore surface whitish.	Tough	White with a thin black line between flesh and the velvety top surface (in cross section).	Tubes.	Forms tiers of small thin brackets. Very common.	Slow to decay. Pore surface remains pale.	Whitish, dry.
14. Daedaleopsis confragosa Fig. 337. (C) (T)	Top at first whitish, later dark reddish or blackish. Pore surface whitish, bruising reddish when fresh.	Tough.	White, later pinkish, then brownish. Bitter taste.	Tubes. Pores elongated radially.	Brackets often strikingly neat and shapely.	Persistent, very dark.	Whitish, dry.
15. Fistulina hepatica Fig. 318, 323. (N) (R)	Top pinkish to reddish. Pore surface whitish to yellowish bruising reddish.	Soft, succulent.	In section, looks remarkably like raw steak. Sour taste.	Tubes not fused together - can be separated from each other.	Season summer-autumn.	Quickly rots away.	At first stains wood rich dark brown. Eventually brown, dry, cubical but not crumbly.
16. Fomes fomentarius Fig. 340. (S)	Top smooth, grey to pale brown. Pore surface pale brown.	Hard outside, tough inside.	Pale brown.	Tubes.	A thick bracket, often markedly hoof-shaped (ie. taller than broad). Rare in South, becoming common northwards.	Persistent.	Yellowish white, mottled, dry with black lines.
17. Ganoderma adspersum and g. applanatum Fig. 320. (M) (S) (R) (T)	Top dark brown to blackish; white rim when growing. Pore surface whitish.	Hard outside, tough inside.	Dark brown, fibrous; silky sheen when torn apart.	Often several clear layers of tubes.	Upper surface and vegetation beneath brackets sometimes covered in cocoa-brown spores, like dust.	Persistent, without white rim.	White, mottled, dry, firm, becoming soft and spongy.

Brackets

Fungus	Top/Underside	Texture	Flesh	Tubes/Gills	Other Distinctive Features	Dead State	Decay Characteristics
18. Ganoderma pfeifferi Fig. 322 (R) (M) (T) (S)	Top dark brown with distinct lacquer-like crust. Pore surface whitish becoming distinctly yellow	Hard outside tough inside.	Like G. adspersum	Like G. adspersum	Spore deposits like G. adspersum. Crust cracks, showing yellow when crushed and melts in match flame. Uncommon.☐ Usually on ☐ Fagus	Persistent without white rim. Yellow colour persists.	Mottled yellow-brown, dry.
19. Ganoderma resinaceum L Fig. 321 (R) (L)	Top like G. pfeifferi but never so hard; often shiny.	Soft with hard crust on top	Pale brown, fibrous.	Tubes pale brown. Usually not clearly in layers.	Like G. pfeifferi☐ but usually on☐ Quercus.	Dry and light in weight, gradually disintegrating.	Eventually a soft, white plastic mass.
20. Heterobasidion annosum Fig. 325 (R) (MD)	Top chestnut to chocolate brown becoming blackish. Rim white in active growth, pore surface whitish.	Tough and leathery.	White.	Tubes.	Nothing very distinctive. Conifer needles, grass etc often grown into fruit body.	Persistent.	Dry, fibrous, pale brown or yellowish with (in longitudinal section) tiny white pockets.
21. Inonotus dryadeus Fig. 327 (R) (M)	Top yellowish to rusty brown. Pore surface whitish.	Soft, moist, spongy.	Rusty brown.	Tubes rusty brown.	Exudes golden, astringent, watery droplets. From spring to autumn. Confined to Quercus.	Blackish, dry, light in weight, pore surface cracked. Slowly disintegrates.	Eventually a soft mass of thin yellowish and white sheets, reminiscent of a mass of cornflakes.
Inonotus hispidus See No. 8.							
22. Laetiporus sulphureus Figs. 345, 346 (R) (M) (T) (S)	Top sulphur yellow to orange sometimes exuding yellow droplets. Pore surface sulphur yellow.	Soft, firm.	Exudes yellowish juice when squeezed. Later dry and white. Pleasant.	Tubes.	Thin, tiered very pretty brackets. From spring to autumn.	Whitish, chalky-looking, light in weight, dry, crumbly slowly disintegrating.	Brown, dry, crumbly, cubical sometimes with chamois leather-like mycelium in cracks.
23. Perenniporia fraxinea (=Fomitopsis cytisina = Fomes fraxineus) Fig. 328 (R) (M)	Top whitish becoming brown to black. Pore surface brownish.	Tough, becoming hard and woody.	Brownish	Brownish, often in distinct layers.	Very uncommon "Ash (fraxinus) Alike"(i.e. ☐ tubes and flesh ☐ concolorous)	Persistent	Brown, dry crumbly, cubical but without mycelial sheets in cracks.

SECTION **3**

Fungus	Top/Underside	Texture	Flesh	Tubes/Gills	Other Distinctive Features	Dead State	Decay Characteristics
24. Rigidoporus ulmarius Fig. 332 (R) (M)	Top whitish becoming yellowish, often with green covering of algae. Pores surface brownish	Tough becoming hard and woody.	Whitish.	Brownish.	"Ulmarius Unlike" (i.e. tubes and ☐ flesh differ in☐ colour)	Persistent.	Brown, dry crumbly, cubical but without mycelial sheets in cracks.
25. Pseudotrametes gibbosa Fig. 351 (T) (C)	Top whitish often with green covering of algae. Pore surface white.	Tough and leathery.	Whitish.	Whitish.	Pores radially elongated. Very Common.	Slowly disintegrates.	White, dry.
Phaeolus schweinitzii See No. 7.							
26. Piptoporus betulinus Fig. 348 (T) (S)	Top smooth, pale brown. Pores whitish	Soft, becoming corky.	White	Tubes.	Confined to Betula. Thick, short stalk sometimes present. Very common. The usual aerial species on birch in the south.	Tuber layer tends to peel away as fruit body begins to disintegrate.	Brown, dry cubical, sometimes with whitish mycelial sheets in cracks.
Pleurotus ostreatus See No. 6.							
27. Polyporus squamosus Fig. 352 (T) (CS)	Top pale brown with conspicuous dark brown scales. Pore surface whitish.	Soft, becoming leathery.	Whitish.	Tubes whitish.	Very shapely and attractive. With short, thick, lateral stalk. Spring to autumn.	Maggot-ridden and slowly disintegrates.	White, dry, spongy or stringy sometimes with tough brown mycelial sheet in cracks.
Rigidoporus ulmarius See No. 24.							

Brackets

SECTION 3

Fungus	Top/Underside	Texture	Flesh	Tubes/Gills	Other Distinctive Features	Dead State	Decay Characteristics
28. Stereum gausapatum Not illustrated (C)	Top hairy, brown. Underside smooth, brown.	Tough; brittle when dry.	Very thin.	No tubes or gills.	Undersurface 'bleeds' (turns red) where scratched (when fresh). Forms a layer on substrate and also well developed brackets.	Buff to grey beneath, never yellow or orange.	Soft, yellowish white, often arranged in several narrow pipes with the grain.
29. Stereum hirsutum Not illustrated (C)	Top hairy, yellowish, underside smooth, yellow becoming brownish.	Tough	Very thin.	No tubes or gills.	Does not 'bleed'. Forms well-developed brackets. Very common	Retains yellow or orange colouration beneath.	Soft, fibrous yellowish white largely confined to sapwood.
30. Stereum rugosum Fig. 383 Also causes a canker-rot. See p. 259 (C)	Top hairy, soon hairless, brown. Underside smooth grey or cream.	Tough, brittle when dry.	Very thin.	No tubes or gills.	'Bleeds' like S. gausapatum. Forms a layer on substrate with very poorly developed brackets. Only on Quercus.	Not distinctive.	Whitish, dry.
31. Stereum sanguinolentum Fig. 350 (C)	Top hairy, grey or brown. Underside smooth grey or pale brown.	Tough	Very thin.	No tubes or gills.	'Bleeds' like S. gausapatum. The only 'bleeding' Stereum on conifers. Forms a layer on substrate with poorly developed brackets.	Not distinctive.	Dry, fibrous, orange-brown with yellow streaks (in longitudinal section).
32. Phellinus tuberculosus (=Ph. pomaceus) Fig. 347 (T)	Top grey, later brownish. Pore surface grey, later dark cinnamon brown.	Hard, woody.	Pale brown.	Tubes.	Confined to Rosaceae.	Persistent and unchanging.	When advanced, crumbly, white, surrounded by purplish brown zone.

← Brackets →

← Layers →

Fungus	Top/Underside	Texture	Flesh	Tubes/Gills	Other Distinctive Features	Dead State	Decay Characteristics
33. Coniophora puteana Fig. 319	Consists of a cream or yellow-brown, smooth or warty sheet of fungal tissue covering the substrate.	Soft.	Very thin.	No tubes or gills	No obvious fruiting structure is produced. Sometimes spreads over adjacent ground or healthy bark.	Not distinctive.	Brown, dry. Longitudinal cracks and very fine transverse cracks.
34. Daldinia concentrica Fig. 339	Fruit bodies are dark reddish brown, later black spheres a few cms across. Smooth with minute warts.	Hard, brittle; charcoal-like.	Black, showing concentric zones when cut through.	No tubes or gills.		Persistent and unchanging.	Small decayed, white patches with black flecks are scattered through healthy wood.
35. Ustulina deusta Figs. 333, 335	Fruit bodies are a black lumpy mass reminiscent of a portion of rough tarmac.	Hard, brittle; charcoal-like	Black.	No tubes or gills.	Thin, circular grey patches with a white margin may be growing close to the black fruit bodies (the asexual state of the fungus)	Persistent and unchanging.	Pale yellowish or greyish with fine, black lines. Dry, brittle. Thin black sheets may line cavities.

← Layers →

← Lumps →

ROTS OF STEMS AND BRANCHES (Top Rots) (See also Section 5)

Fungi: *Various species, mostly bracket and gill fungi*

Damage Type: Wood decay.

Symptoms & Diagnosis: Areas of exposed wood (i.e. no longer covered by bark) on stem or limbs are evidently decayed, or a cavity has formed at such a wound; or fruit bodies of known wood-rotting fungi form on bark or exposed wood on limbs or stem (Figs.320,322,336–352); or a stem or limb snaps.
Confirmation: When tested with a knife, chisel or other instrument, decayed wood is usually palpably softer or more readily picked out, broken or cut than sound wood.
Additional indicators are given in Section 5, Decay and Safety.
Caution: Even perfectly sound limbs and stems will break if the wind is strong enough. In calm weather, apparently sound limbs may fall – a phenomenon known as 'summer branch drop' (see p.178).

Status: Top rots are common and widespread, particularly where large branches have been cut off or have broken out.

Significance: Ultimately may render the tree or limb unsafe; in top rots where sapwood or bark is also killed (e.g. Silver leaf, p.252), the decay may be of less concern than the bark killing. The identification and the significance of the various top-rotting fungi is dealt with above, at the end of the entry on Rots of Roots and Butt.

Host Trees: It is doubtful if any species is immune from attack by every one of our many top-rotting fungi but some are particularly prone to such disease, Fagus sylvatica among them. See also Section 2.

Infection & Development: Airborne spores are released by fruit bodies borne on decayed parts of standing or fallen trees or on infected timber left in the open. For infection, most fungal species probably need wood unprotected by living bark and even then infection is more likely if the end grain is exposed, as in pruning wounds, or where the wood is not only exposed but also torn. Some top rots result from the development of endophytic fungi (see Strip canker of beech, p.114 and Comment under Betula on p.28).
The fungus slowly spreads through the wood, converting the tissues to its own substance and to energy for growth. The manifestation of this process is the decay of the wood. Once the fungus has exhausted all the materials it is capable of utilizing it dies out in that portion of wood. The decayed wood may be further broken down by other micro-organisms, arthropods and birds so that finally a cavity is formed. The ultimate extent of the decay is restricted by various anatomical, chemical and physiological characteristics of the wood (see Shigo, 1986).

Control: *Therapy:* None available: the excavation of decayed areas will not arrest the spread of the fungus and can result in a greater final volume of decayed wood than might otherwise have been the case; at the same time, the surgery itself weakens the limb or stem involved. Once a wound has callused over, further spread of the decay ceases or slows markedly.
Where tree safety is the prime consideration: Tree hazard assessment, the detection of decay and available control measures are discussed briefly in Section 5 in this book and in detail in Lonsdale (1999).

SECTION 3

Prevention: Protect bark of limbs and stems from mechanical and fire damage; prune as little as possible; to avoid creating large pruning wounds, remove potentially troublesome branches early; follow the pruning guidelines in British Standard No. 3998 (Anon, 1989). Also see Preventing Decay on p.296.

No wound treatment has yet been found which protects large wounds against decay organisms for a usefully protracted period, while wounds which will callus over in a few years require no treatment. Furthermore, some wound paints can make decay more likely, not less. For these reasons, wounds should be left bare except where protection against Bacterial canker of cherry or Silver leaf is required (and even these diseases can be guarded against by pruning only in summer: see pp.72 and 254).

Remarks: The effect on the tree of the various top-rotting fungi varies hugely depending on the species of fungus and tree involved. The identity of the fungus can therefore often be invaluable in prognosis and in deciding on appropriate control measures. Identification from the decay alone usually requires laboratory techniques but if fruit bodies are associated with the symptomatic tree, the fungus can often be identified from these. A means for identifying in the field all of the species likely to be encountered and notes on their significance are presented on pp 226-234.

Further Reading:
Lonsdale, (1999); Mattheck & Breloer (1994).

Fig. 336 *Coriolus versicolor*

Fig. 337 *Daedaleopsis confragosa (= Trametes rubescens)*

Fig. 338 *Bjerkandera adusta*

Fig. 339 *Daldinia concentrica* (Cramp Balls)

Fig. 340 *Fomes fomentarius.* See Comment p.28

Fig. 341 *Inonotus hispidus* (young fruit body)

SECTION 3

Fig. 342 *Inonotus obliquus.* See Comment p.28

Fig. 343 *Inonotus hispidus* (old fruit body)

Fig. 344 *Flammulina velutipes*

Fig. 345 *Laetiporus sulphureus* (Chicken of the Woods) (young fruit bodies)

Fig. 346 *Laetiporus sulphureus* (old fruit bodies)

Fig. 347 *Phellinus tuberculosus*

Fig. 348 *Piptoporus betulinus*
(Birch polypore). See Comment p.28

Fig. 349 *Pleurotus ostreatus*
(Oyster mushroom)

Fig. 350 *Stereum sanguinolentum*

3

SECTION

Fig. 351 *Pseudotrametes gibbosa*

Fig. 352 *Polyporus squamosus*
(Dryad's Saddle)

RUST OF BIRCH, POPLAR AND WILLOW LEAVES

Fungi: *Melampsoridium betulinum; several Melampsora species*

Fig. 353

Fig. 354

SECTION 3

Damage Type: Leaf killing and premature defoliation. (Some Melampsora species cause cankers on current willow shoots.)

Symptoms & Diagnosis: In late summer, leaves are conspicuously orange or yellow due to the formation on the underside of the leaf (in Salix, on both sides) of slightly raised, powder- (spore) filled pustules (Fig.353, on Populus) coincident with discoloured flecks on the other leaf surface (Fig.354, on Populus). The pustules are individually small but may be very numerous. Later, inconspicuous, yellow or brown, flattened, blister-like areas appear on either leaf surface. Affected leaves die and fall prematurely (or, in poplar, after hanging withered for some weeks).
Additional indicators: On larger trees, the damage is worse in or confined to the lower crown (Fig.355, on Populus). Similar pustules occur on petioles and soft shoots. There is an abundance of the alternate host of the fungus close to the affected tree (see Infection & Development below).
Caution: Taphrina populina (p.197)

Fig. 355

causes yellow patches on the underside of poplar leaves but these are not raised, powdery pustules. Other diseases also cause leaf spotting and premature defoliation of birch (p.163), poplar (pp.165,245) and willow (pp.167,245).

Status: Common and widespread.

Significance: Startling in appearance but of no consequence on established amenity trees. The defoliation is known to cause a marked growth reduction in poplars, so severe outbreaks on newly planted trees may delay their establishment. See also Remarks. Can require control in nurseries, osier beds and biomass plantations.

Tree susceptibility and Resistance: The many poplar and willow species, hybrids, cultivars and clones grown in this country vary widely in their susceptibility to the numerous Melampsora species which attack them here. None of them, however, nor any of the birches grown here, are so frequently or severely attacked that the diseases need to be taken into account when choosing trees for amenity planting.

Infection & Development: In summer, the orange or yellow airborne spores (urediniospores – Fig.353) rapidly spread the disease through the infected trees; during winter the fungi survive in the fallen leaves. In spring, in most species, basidiospores are produced from the overwintered leaves, but these are incapable of infecting the species of tree from which the leaves came. Instead, they infect some altogether different plant (according to fungal species, Allium, Arum, Mercurialis, Larix or Pinus in the case of Poplar rusts; Allium, Euonymus or Larix for Willow rusts; and Larix for the birch rust). Later in the year, airborne spores (aeciospores) of a type which can infect the original tree species are produced on these 'alternate' hosts and once again give rise to the disease on poplar, willow or birch leaves.

This alternation of the fungus between two unrelated host plants is the classic 'text-book', full life cycle of a rust fungus. However, several, if not all, of the rusts under consideration here also produce in spring urediniospores from the overwintered leaves; or some urediniospores produced in late summer overwinter under the protection of bud scales on the tree. In either event, the first poplar, birch or willow leaves to emerge in spring are often infected by these spores direct, so the disease can occur again in the following year without the involvement of the alternate host.

Control: None required in amenity trees.

Remarks: By interfering with the normal winter hardening process, the premature defoliation sometimes results in winter cold killing of current shoots, damage which may be increased by secondary attacks of weak pathogens such as Cytospora on willow, Discosporium on poplar and Melanconium on birch, but recovery is rapid if cultural conditions are favourable.

RUST OF BOX LEAVES

Fungus: *Puccinia buxi*

Damage Type: Leaf thickening; shoot tip killing.

Symptoms & Diagnosis: During summer, dark, thickened patches appear on the current year's leaves (Fig.356). In late summer, some shoot tips may be dead.
Confirmation: During summer, purplish brown, blister-like pustules (containing the infective spores) may be seen on the thickened leaf spots, on either side of the leaf.

Status: Quite common, especially in Scotland.

Significance: A little unsightly but no threat to the general health of the plant.

Host Trees: Buxus sempervirens. Also recorded on B. balearica on the Continent.

Resistant Trees: No information.

Infection & Development: The fungus overwinters in infected box leaves which are still attached to the tree. In spring (and probably during the whole growing season in suitably moist conditions), spores produced in purplish brown pustules on dark leaf spots on either side of these leaves infect the tender, newly developing leaves. As the fungus spreads within the leaf, the palisade cells are stimulated to elongate unnaturally, and the spongey parenchyma to swell. In consequence, the infected part of the leaf becomes swollen. Young, succulent shoots are also infected and sometimes killed. The newly infected leaves carry the fungus through the next winter. It is not clear from the literature consulted whether the leaves die or fall after releasing the spores.

Control: None required.

Remarks: The literature on this disease is sparse.
Unlike the classic rusts of the textbooks, this disease involves no alternate host plant. The fungus is also peculiar in that the teliospores themselves infect the leaves (in most rusts, the teliospores produce basidiospores, and it is these which are the infective agents).

Fig. 356

SECTION 3

RUSTS OF HAWTHORN and ROWAN LEAVES RUST OF JUNIPER STEMS

Fungi: *Gymnosporangium species*

Fig. 357

Fig. 358

Damage Type: On hawthorn and rowan, leaf killing; on juniper, distortion of branches and some branch killing.

Symptoms & Diagnosis: (i) On hawthorn and rowan: In late summer, reddish, orange or brown thickened spots appear on leaves; on hawthorn, some current shoots are strongly curved and thickened and some haws are distorted. From the underside of affected leaves and from damaged shoots and fruit protrude clusters of peg- or horn-like, cylindrical or conical, yellowish or brownish outgrowths, 1–5 mm high (Fig.357). These rupture to release the brown, powdery contents (aeciospores). Later the leaves turn brown, curl and shrivel. (ii) On juniper: Portions several centimetres long of some twigs and branches are abnormally thick, the thick portion tapering at each end (spindle shaped). In spring, scattered on these swellings, are conspicuous, yellowish or brownish cushions or 5–10 mm long, tongue-like outgrowths (teliospores), gelatinous in wet weather, horny-hard when dry (Fig.358). Swollen branches are usually alive; on dead swellings, no teliospores are produced.

Additional indicators: A juniper bush is in the vicinity of infected rowans or hawthorns, or vice versa.

Status: Frequent where junipers grow near hawthorns or rowans.

Significance: An intriguing curiosity when seen for the first time but of no importance in the growing of the trees.

Host Trees: Sorbus aucuparia; Crataegus monogyna and C. laevigata; Juniperus communis.

Infection & Development: In autumn, windblown spores are produced from infected hawthorn and rowan tissues (Fig.357). These aeciospores are incapable of reinfecting the plants which bore them, and infect instead current shoots of junipers. The fungus becomes established permanently in these shoots, locally stimulating their annual radial growth. In spring each year, spores (teliospores) are produced in gelatinous masses from the swollen stems (Fig.358) and in turn, in humid weather, produce airborne basidiospores. The basidiospores cannot infect juniper but are carried away to infect hawthorn or (depending on the fungal species) rowan again on which, in late summer, aeciospores are once more produced.

Control: None required.

Remarks: Only occasionally, and after some years, do the swollen juniper branches die.

In the classic 'text book' rust diseases, spores produced during the summer reinfect the plant on which the spores (urediniospores) are produced. In these Gymnosporangium diseases, this phase is absent.

The connection between the disease on hawthorn and that on juniper was confirmed in 1867, in Denmark.

SCAB DISEASES
(see also Leaf Spots and blotches, p161)

Fungi: *Venturia species* (anamorphs: *Spilocaea, Pollaccia, & Fusicladium species*)

Damage Type: Killing of leaves and shoots; in some species, defoliation.

Symptoms & Diagnosis: *In Malus, Pyrus and Sorbus:* during summer, varying degrees of leaf fall occur (Fig.359, on Malus); diseased leaves bear brown or olive-green spots and blotches and some are yellowed and distorted (Fig.360, on Malus); discoloured and green leaves bear diffuse, sooty patches consisting of a network of very fine black threads; similar spots appear on fruits, which then develop roughened, corky patches and cracks (Fig.361), and on flower parts.

In Populus and Salix (Fig.364 on Populus; Figs.362,363 on Salix): black spots and blotches appear on leaves; whole leaves die and shrivel but remain firmly attached; black lesions form on shoots around the base of petioles; shoot tips and small side shoots blacken and curve over and the attached leaves die and hang conspicuously shrivelled on the tree.

Confirmation: A thin, velvety, olive-green to almost black mould forms on leaf (Fig.360), shoot and fruit lesions (usually plentiful on Malus, Pyrus and Sorbus, and on surrounding live tissue; often on few of the lesions and rarely on shoots of Populus and Salix).

Additional indicators: Outbreaks follow cool, wet weather.

Caution: On Salix, closely similar symptoms are caused by a much less common fungus, Glomerella miyabeana. (See Rose, 1989b).

Status: Widespread. Very common on Malus and Salix; occasional on Populus; infrequent on Sorbus; probably uncommon on ornamental pears but common on fruiting varieties.

Significance: Very disfiguring and can make replacement of very susceptible

SECTION 3

trees, especially crabs, desirable. One of the most troublesome diseases of apples and pears grown for fruit.

Host Trees: Malus (the crabs vary widely in susceptibility – some of the purple-leaved varieties are particularly susceptible) and Sorbus (Venturia inaequalis/ Spilocaea pomi); Populus, esp. aspen, (V. tremulae/P. radiosa and V. populina/ P. elegans); Pyrus (V. pirina/F. pyorum); Salix, esp. S. matsudana 'Tortuosa', S. alba vitellina and S. fragilis (V. saliciperda/P. saliciperda).

Resistant Trees: Salix alba caerulea, S. babylonica, S. pentandra, S. purpurea. Of the poplars, only aspen seems particularly susceptible. There is no useful information on the relative susceptibility of the Sorbus species.

Infection & Development: The fungus overwinters on fallen leaves (except in Salix) and in shoot lesions; in spring, spores from these infect young leaves, shoots, sepals and fruit; the fungus also grows down petioles into shoots; spores produced on the new lesions give rise to more infections in suitably wet summer weather.

Fig. 360

Fig. 361

Fig. 362

Fig. 359

Fig. 363

Fig. 364

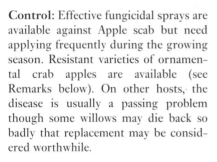

SECTION 3

Control: Effective fungicidal sprays are available against Apple scab but need applying frequently during the growing season. Resistant varieties of ornamental crab apples are available (see Remarks below). On other hosts, the disease is usually a passing problem though some willows may die back so badly that replacement may be considered worthwhile.

Remarks: In the first edition of this book, lists of crab apples deemed in the USA to be resistant and susceptible to Apple scab (and Powdery mildew) were given. Since then the frequent and quite severe damage suffered here by one of the species stated to be resistant, namely Malus tschonoskii, and the good health of one said to be susceptible, M. 'Charlottae', throws doubt on these lists' validity for this country. They have therefore been deleted. Information from collections of Crab apples here on this matter would be useful and relatively easy to collect.

Apple scab was described from Europe in the early 19th century.

Further Reading: (on Salix): Rose, 1989b.

SCALE INSECTS

Damage Type: General debilitation of tree; dieback.

Symptoms & Diagnosis: (Figs.365–368 and others referred to in the text) Trees appear sickly, grow poorly and perhaps die back.
Confirmation: Large numbers of brown or greyish protruberances (scale insects) are present on twigs, stems or foliage. Individual scales are up to 6–7mm diameter, but often smaller, and may be globular, elongate, circular and flattened or shaped like a mussel shell (often rather similar to galls, see p.132).
Additional indicators: Large or small amounts of a whitish waxy wool are associated with some species (see Pulvinaria p.149, Cryptococcus p.120). Sticky Honeydew is produced copiously by some species, but not others and is later blackened by Sooty moulds (see p.144).

Significance: Some scale insects have a debilitating effect on the host which may lead to dieback in extreme cases, while other species can be present in large numbers with no apparent damage to the host tree (see species accounts below for further details).

Host Trees: Scale insects affect a wide variety of trees (see below). Conifers are less affected than broadleaves but Carulaspis spp. and Parthenolecanium pomeranicum are damaging to Cupressaceae and Taxus respectively. A few particularly notable species have been given a separate entry.

Control: On deciduous trees tar oil winter wash applied during the dormant season will reduce high populations of scale insects. Otherwise, the young mobile stages are most susceptible to a persistent contact insecticide applied when the eggs are hatching (see examples below). [Take note of current Pesticides Regulations – see p.288] Scrubbing the bark surface with a mild detergent solution will remove infestations from bark.

Further Reading: Detailed accounts of some species are given by Alford (1991).

SOME COMMON SCALES OCCURRING ON AMENITY TREES

Brief descriptions are given below of some of the scales encountered most frequently on amenity trees. Several of these scale insects are polyphagous, occurring on a wide range of hosts, but are listed here under the genus on which the author has most often seen the species.

FAGUS:
Cryptococcus fagisuga,
Felted beech scale (coccus) (see p.120).

JUNIPERUS:
Juniper scale
Carulaspis juniperi; C. minima
(indistinguishable from each
other in the field)

Fig. 365

Affected plants appear discoloured and unthrifty (Fig.365); the foliage has a dull appearance. On close examination it is seen to be 'spotted' with scales, 1.0–1.5mm diameter, pale with a yellowish spot off centre; these occur on berries as well as foliage (Fig.366). Most often reported on junipers but these scales affect a wide range of genera and species in the family Cupressaceae and they are also found on Sequoiadendron. Control with insecticide in June or July when the nymphs first appear. [Take note of current Pesticides Regulations – see p.288] Overwintering scales are often predated by birds.

Fig. 366

SECTION 3

PRUNUS AND OTHER BROADLEAVES:
Brown scale
Parthenolecanium corni

Large brown oval scales, 4–6mm long and similar in appearance to P. pomeranicum (Fig.367) are found on stems from autumn to spring. Their presence may lead to premature leaf fall, but otherwise they cause little damage on rosaceous trees. The nymphs feed on the leaves during the summer. Most P. corni are parthenogenetic and eggs are laid beneath the scales in early summer. P. corni occurs on Prunus, Magnolia and Malus; in Europe it is also known on Robinia. Control is not usually necessary but if required tar oil winter wash will be effective in the dormant season. Alternatively, apply a contact insecticide when the 'crawlers' hatch in the summer and move onto the leaves to commence feeding. [Take note of current Pesticides Regulations – see p.288]

SALIX:
Willow scale
Chionaspis salicis

The whitish appearance of the stems is caused by encrustations of both male and female scales on the bark and is most noticeable during the winter. In winter eggs below the yellowish white female scales (1.5–2.3mm long) can be squashed with the thumb-nail to give a blood-like smear. The males are smaller (0.5–1.0mm) and more elongate. The eggs hatch in May when the red nymphs gather in noticeable clusters on the bark.
C. salicis also occurs commonly on Alnus and Fraxinus. It is not thought to seriously affect the host but some 'dimpling' of the bark has been observed, particularly on young stems and branches. Control is not recommended.

Fig. 367

TAXUS:
Yew scale
Parthenolecanium pomeranicum

Similar in appearance to P. corni (above) but occurs mainly on the foliage (Fig.367). This scale insect excretes copious quantities of honeydew which spoils the foliage when it is colonised by Sooty moulds (see p.144). Such damage on yew hedges can be controlled with a contact insecticide in July. [Take note of current Pesticides Regulations – see p.288]

TILIA:
[Horse chestnut scale (see p.149), *Pulvinaria regalis*.]

VARIOUS BROADLEAVES:
Nut scale
Eulecanium tiliae

This scale is seen as chestnut, greyish or light brown convex protruberances, 5–6mm diameter, on twigs and small branches (Fig.368). Some dieback of the affected parts may occur, on young trees in particular. Found on a wide variety of broadleaves including Alnus, Quercus, Tilia, Carpinus and rosaceous trees but is rarely reported causing serious damage. The life cycle and control are similar to Parthenolecanium corni (see above).

Fig. 368

SILVER FIR WOOLLY APHID (ADELGID)   Insect: *Adelges nordmannianae*

SECTION 3

Damage Type: Needle distortion; shoot dieback.

Symptoms & Diagnosis: On the current shoots from May onwards the needles are either twisted, distorted and shortened, making some shoots look rather like bottle brushes (Fig.369) or needles are dropping and/or shoots are dying back (Fig.370).
Confirmation: Plentiful honeydew (later colonised by Sooty moulds – see p.144) is produced by small black adelgids (aphids) on the needles. Sparse greyish waxy wool, which may be found on older stems or trunks at any time, also indicates the presence of this adelgid (Fig.371).
Caution: Similar needle distortion can be caused by hormone weed killers.

Status: Occurs throughout Britain wherever its host is grown.

Significance: Disfiguring on amenity trees where, if attacks persist, the growth is inhibited and the crown will be reduced. A serious problem in Christmas tree plantations.

Host Trees: Abies alba, A. cilicica, A. nordmanniana and other Eurasian Silver firs. An alternate generation causes galls on Picea orientalis (see below).

Fig. 369

Life Cycle: Nymphs of A. nordmanni-anae overwinter at the needle bases. They commence feeding in March and orange-brown eggs laid in mid-April hatch as the buds burst in late April or early May. The crawlers of this generation move to the new shoots to feed; winged forms migrate to P. orientalis where small pineapple-like galls develop in place of shoots the next summer.

Control: Only recommended in Christmas tree plantations, where a permitted pyrethroid insecticide is applied as a high volume spray between early November and the end of February. [Take note of current Pesticides Regulations – see p.288]

Further Reading: Carter (1971), Carter & Winter (1998).

Fig. 370

Fig. 371

SILVER LEAF	Fungus: *Chondrostereum purpureum*
	(= Stereum purpureum)

Damage Type: Disruption of the tree's metabolism and water transporting system.

Symptoms & Diagnosis: All the leaves on one or several branches have assumed a leaden or silvery coloration (Fig.372); some may have wilted and died and some fallen prematurely. Some branches may be dead and leafless.

Additional indicators: There are pruning or other wounds on or close to branches showing symptoms. *On Malus:* outer layers of bark on some affected branches are blistering and peeling away (as in Fig.12, p.44).

Fig. 372

Fig. 373

Fig. 374

Fig. 375

Confirmation: Wood of affected branches shows a dark brown stain in cross-section (perhaps only at the branch base or near a wound). A narrow strip of dead bark may extend back from the dead branch into the live part (Fig.373, on Rhododendron). Dead branches or associated wounds may bear tiers of small fungal brackets, off-white and woolly above, the underside smooth and brown when dry (Fig.375), purplish when fresh (Fig.374). If a portion of the stained wood is wrapped in damp paper for a week or so, the stained area produces a dense, white mycelium.

Caution: Leaf silvering can result from excessive fertiliser application ('False silver leaf') or Bacterial canker (p.70). A few hosts (e.g. Rhododendron and Laburnum – but see Hosts below), die back without silvering. See also Ligustrum Comment p.42.

Status: Both fungus and disease are common and widespread.

Significance: A progressive and often fatal disease but the fungus also commonly occurs asymptomatically on many tree species.

Susceptible Trees: Many Rosaceae, particularly Prunus (Fig.372) and

Malus species; also Rhododendron (Fig.373). It is associated with and probably the cause of a dieback in Eucalyptus. The authoritative T. R. Peace (1962) states that it can cause severe dieback in Laburnum.

Resistant Trees: Conifers are immune. Some fruit tree varieties are relatively resistant. The relatively few records of the disease on the very common Prunus laurocerasus suggests it may be fairly resistant. The fungus causes a sapwood decay in many other species (e.g. Salix, Populus) but does not kill branches.

Infection & Development: Airborne spores produced from fruit bodies on infected wood infect freshly exposed, live sapwood, especially end-grain. Toxins produced by the fungus during its growth in the sapwood trigger anatomical changes in the leaves – manifested as silvering – and the production of gums and tyloses which block the water-conducting vessels. Branches die from direct fungal invasion, water starvation and poisoning. The fungus can grow via natural root grafts into adjacent trees and kill them from the bottom upwards.

Control: *Therapy:* Take no action unless dieback occurs or the whole tree is silvered, as silvered branches may recover spontaneously, especially in Malus. To arrest the spread of the disease, cut dead and dying branches back into unstained wood and apply wound paint (if necessary – see below). Entirely silvered trees are unlikely to recover and should be removed.

Prevention: Most infections take place through pruning wounds but infection is unlikely in gum-producing species (Prunus) if these are pruned in June, July or August. Otherwise, treat fresh wounds with an effective wound paint (e.g. octhilinone, Trichoderma). [Take note of current Pesticides Regulations – see p.288]

Burn, bury or remove from the vicinity of susceptible species all dead or dying wood in standing trees or lying on the ground which bears or may later produce fruit bodies of the fungus.

Remarks: The fungus is probably indigenous. The cause of the disease was determined in the early 20th century, in England.

SMALL CEDAR APHID   Insect: *Cedrobium laportei*

Damage Type: Defoliation; dieback.

Symptoms & Diagnosis: In late spring or early summer a rapid loss of the old needles causes an unthrifty appearance of the whole tree and some shoots may die (Fig.376).
Confirmation: Honeydew drips copiously from the tree and Sooty moulds occur subsequently wherever the honeydew has accumulated (see p.144). Search between the scales on the dwarf shoots in May or June for small pale brown and grey aphids (1.5–2.0mm long), with or without wings; these are very difficult to find.

Status: Distribution incompletely known. Recorded from southern England as far north as Lincolnshire and in Dundee.

Significance: Severe defoliation, from which the badly affected parts of a tree may not recover, leads to a loss of the shading effect.

Host Trees: Cedrus atlantica, C. deodara, C. libani.

Life Cycle: This is incompletely known in Britain. Both apterous (wingless) and alate (winged) viviparous females (i.e. aphids that give birth to live young) occur in May and June; the winged form has also been found in early July. Wingless aphids can also be found until October. This species perhaps overwinters as an egg.

Control: Chemical control is not recommended because of the practical and legal problems involved in spraying large amenity trees with pesticides.

Further Reading: Carter and Maslen (1982).

Fig. 376

Remarks: C. laportei has spread northwards rapidly through Europe since 1967 and was first found in Britain at Kew in 1974.

SECTION **3**

SMALL POPLAR LONGHORN BEETLE	Insect: *Saperda populnea*

Damage Type: Swellings (galls) on one- or two-year-old stems.

Symptoms & Diagnosis: Swellings, usually about 6–12mm diameter and rather like an old-fashioned plumber's joint, develop on one or two year old stems and branches (Fig.377). There may be several of these galls close together on the same stem, sometimes as little as 25mm apart.
Confirmation: A brown horseshoe-shaped scar is normally present on each gall. The gall and the adjacent stem are hollow and filled with light orange bore dust. A chamber, plugged with wood fibre, may be present in the wood immediately above or below the gall in which there may be a larva or pupa (Fig.378). Circular emergence holes 3–4mm diameter appear from late May onwards leading from these pupal chambers.

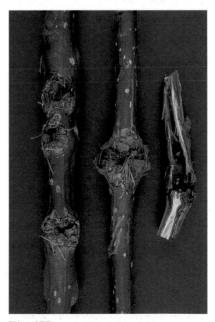

Fig. 377

Status: Widespread and sometimes common.

Significance: Affected stems are liable to break where the gall is formed and before the wound occludes.

Host Trees: Populus tremula is the preferred host, but galls occur on many poplars including modern hybrids. Also on Salix caprea, S. alba, S. fragilis, S. viminalis and exceptionally on Betula, Corylus or Fraxinus.

Life Cycle: This spans one or two years. Data from Britain is inconclusive, but in continental Europe the life cycle ranges from an annual generation in the south to one every two years in Sweden. Adult beetles in May and June lay eggs into small horseshoe-shaped punctures in the twigs. The eggs hatch in about 2 weeks and the larvae tunnel into the sapwood, causing galls to develop. They pupate in spring in a special chamber filled with wood fibres and located in the stem immediately above or below the gall. NOTE: Sometimes S.populnea galls form after the female beetle lays eggs into the stem but without the larvae developing; these galls remain empty and a little smaller than occupied examples.

Control: Not usually justified. If serious damage recurs the use of an insecticide when the eggs hatch may give some control. Brushing linseed oil onto

Fig. 378

the horseshoe scars has been tried with some success in Sweden.

Remarks: Smaller pear-shaped galls without the depression and puncture hole are caused by the clearwing moth Synanthedon flaviventris which is only found in southern England. Galls on larger diameter stems, most often at the base, are due to the weevil Cryptorhynchus lapathi which causes a problem in osier beds, but apparently not on amenity trees.

SOOTY BARK DISEASE OF SYCAMORE

Fungus: *Cryptostroma corticale*

Fig. 379

Fig. 380

Damage Type: Not fully elucidated. Probably a combination of poisoning and interference with water transport.

Symptoms & Diagnosis: *Syndrome A:* The foliage of parts or the whole of a tree's crown has wilted; the wilted leaves are dry and faded and cupped upwards. Shrivelled dead, brown leaves hang firmly attached to the twigs (Fig.379). *Syndrome B:* In spring, the whole of or parts of a tree have failed to come into leaf and some parts bear abnormally small, sparse leaves.
Confirmation: (i) *Wood stain:* If an affected branch is cut across or a tree cut down before it is completely dead, patches of green-, yellow- or brown-stained wood, often with a narrow, dark margin, will be found (Fig.380), appearing as long, discoloured bands if viewed in longitudinal section. If the stain involves the outermost wood, the bark will be found to be dead and perhaps showing (ii) sooty bark: patches or strips of bark, roughly proportional in extent to the crown symptoms, may be dead, puffy and blistered (detectable by finger pressure). Some outer bark may have flaked off exposing a layer of black, soot-like powder (spores) within the thickness of the dead bark (Fig.381). (iii) Black bark: late in the development of the disease, large areas of smooth, black, dead bark may be present, perhaps flaking off.
Caution: Severe drought on its own can kill sycamore bark. The stain caused by C. corticale disappears once the wood is dead. Sycamore wood stains readily from other causes.

Status: The fungus is widespread on dead wood but requires long, hot, dry summers to cause disease. Outbreaks

SECTION **3**

are, therefore, sporadic and concentrated in the southern half of the country.

Significance: Numbers of deaths, even among large trees, can be alarming and the local impact of the disease considerable, but the short-lived and geographically restricted nature of epidemics limits the effects of damage in the long term to tolerable levels.

Host Trees: The disease is almost confined to Acer pseudoplatanus, but has occasionally been seen on other maples. The fungus has been found growing saprophytically on Aesculus hippocastanum in this country and on other broadleaves abroad.

Infection & Development: The fungus is an endophyte, that is to say it can exist in a latent state in the wood of healthy trees for many years, causing no symptoms. However, if high temperatures and prolonged dry weather prevail it can spread rapidly in the wood and cause the disease. The extent of this spread is evidenced by the stain in the wood but external symptoms do not appear until the outermost sapwood rings are invaded (Fig.380). Then wilting sets in, bark dies, and the fungus spreads through it to form its very extensive fruiting structures. On maturity, these break open as the thin, blistered outer bark flakes off and the exposed powdery spores drift away in the air. The exposed black inner bark weathers to a smooth surface and eventually falls away.

Fig. 381

Control: *Therapy:* None available.
Prevention: Not possible, but the impact of outbreaks will be lessened if, in the southern half of the country, sycamore, where used, is mixed with other species.

Remarks: London was the first recorded location in Great Britain both for the fungus, in 1945, and the disease, in 1948. It was possibly introduced from North America as it was first described from maple logs in London, Ontario in 1889.

Further Reading: Young, 1978.

STEREUM CANKER-ROT

Fungus: *Stereum rugosum*

Damage Type: Bark killing and wood decay.

Symptoms & Diagnosis: On the stem (or a limb) at any height above the ground, a roughly elliptical, depressed area is evident, a few centimetres or as much as 1.5 metres high and extending part or most of the way round (Fig.382). Viewed face on, the sides of the depression (canker) may bulge out beyond the normal line of the stem (Fig.383). The canker may clearly consist of a number of ridges arranged concentrically around a twig or branch or their remains. The bark covering the canker is dead and some may have fallen off.

The stem may have died above or snapped at the canker.

Confirmation: The wood beneath the canker is decayed (Fig.384); fruit bodies of Stereum rugosum (hard,

Fig. 383

Fig. 382

Fig. 384

SECTION **3**

buff-coloured, coin-sized, appressed patches, often numerous, which, when fresh, appear to bleed if scratched) are present on the dead bark (Figs.382,383).

Status: An uncommon disease though locally frequent. As the fungus is a common saprophyte on stumps and fallen branches, its role on living trees when not clearly associated with perennating cankers is uncertain; it probably kills feeble twigs and observations suggest that it may sometimes cause a true dieback of otherwise apparently healthy branches.

Significance: Ornamentals can tolerate the cankers for years but in the end the stem is likely to snap. The canker-rot will reduce the timber value of a tree.

Host Trees: The disease has most often been recorded on Quercus rubra (Fig.383), less often on Q. robur, rarely on Fagus sylvatica, Nyssa sylvatica (Fig.382), Prunus padus and Sorbus aria, and abroad on other species; on Rhododendron it may cause a dieback. As a saprophyte the fungus is common on a very wide range of broadleaves.

Infection & Development: Airborne spores infect twigs or branches, often through pruning wounds; whether unwounded twigs are susceptible is not known but most cankers arise on the lower stem, suggesting an association with pruning wounds or supressed branches.

The fungus grows down to the base of the branch or twig and then, during the dormant season, radially through the cambium and phloem of the stem to kill an elliptical patch of bark. In the following growing season, its spread is arrested and the lesion begins to callus over. In the next dormant season, however, the fungus kills the callus and enlarges the lesion. This alternate killing and callusing continues each year and results in the formation of a so-called 'target canker'. The fungus also penetrates and rots the sapwood and heartwood (Fig.384). The canker-rot may spread indefinitely until either the stem is girdled and dies or it snaps. Fruit bodies are produced on the freshly killed bark.

Control: *Therapy:* In theory, infected tissues could be excised, but the resultant cavity would itself weaken the stem and be vulnerable to further infections by wood-rotting fungi.
Prevention: The disease is too rare and too poorly understood for preventative measures to be warranted or suggested at present.

Remarks: The disease was first described in 1901, in England, when the fungus was shown to be pathogenic to oak. The perennating nature of artificially induced cankers has not yet been convincingly demonstrated experimentally, however.

VERTICILLIUM WILT	Fungi: *Verticillium dahliae, occasionally V. albo-atrum*

Fig. 385

Fig. 386

Damage Type: Disruption of water flow within the tree; poisoning.

Symptoms & Diagnosis: The foliage of one or a few branches has wilted and died (Fig.386 on Golden catalpa) or in spring a few branches have flushed weakly or failed to flush. Wilted leaves are falling prematurely (Fig.385). The tree may seem otherwise healthy or some dead branches may be present.
Additional indicators: Elongated patches of dead bark on the stem; bark cracks; gum exuding from the bark.

Fig. 387

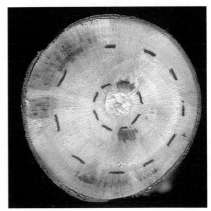

Fig. 388

SECTION 3

Confirmation: Cross sections of affect-ed branches show green or olive-brown spots or blotches (streaks in longitudinal section) following the current and perhaps earlier annual xylem rings (Fig.388. Red ink marks indicate annual rings) (see also Remarks below). Streaks or sheets of the same stain may be evident immediately under the bark of diseased parts (Fig.387 shows a diseased and a healthy branchlet).
Caution: This characteristic staining may be present only towards branch bases or in the main stem, and in Magnolia may not occur at all. Green-stained wood is common in some Acer species in association with injuries.

Status: Common in some ornamental tree nurseries and outplanted trees from these. Occasional otherwise.

Significance: Often fatal, especially to recently planted trees. Trees can recover after the death of some branches, though the disease may recur after an interval of some years.

Host Trees: Recorded on many species of woody and herbaceous plants but on trees only common on Acer (A. platanoides in particular – Fig.385). Catalpa, Cercis, Koelreuteria and Tilia are also rather susceptible. Some discrepancies between different authors' lists of susceptible species are due to variations in the fungus and because some tree genera contain both susceptible and resistant species. In Holland and the USA the disease affects species of Fraxinus: see Comment (3) under Fraxinus on p. 37.

Resistant Trees: Conifers are immune. The following are immune or very resistant: Alnus, Betula, Carpinus, Castanea, Cercidyphyllum, Cornus, Crataegus, Cydonia, Eucalyptus, Fagus, Ginkgo, Gleditsia, Ilex, Juglans, Liquidambar, Malus (despite some reports to the contrary), Morus, Platanus, Populus, Prunus, Pyrus, Quercus, Salix, Sorbus (except S. torminalis), Zelkova.

Infection & Development: The fungus persists in the soil and in roots of symptomless woody and herbaceous plants for years. It invades via intact or damaged roots and spreads through the tree in the xylem vessels. The wilting and death are probably the combined effect of direct poisoning by the fungus and the interruption of the passage of water through the tree which results from the tree's defensive reaction. Infected leaves reinfect soil. If the tree survives one year, reinfection from the soil is necessary for a recurrence of the disease as the fungus often does not spread from annual xylem ring to ring. Stresses such as transplanting, drought or deicing salt damage can worsen the disease.

Control: *Therapy:* Heavy watering and applications of nitrogenous fertilizers to recently wilted trees may, by stimulating the production of new, uninfected wood, enable the tree to survive.
Caution: Use ammonium fertilizers, NOT nitrates, which can make matters worse.
Prevention: Use planting stock from disease-free nurseries; burn debris from diseased plants; if a tree dies, replant with resistant species.

Remarks: In the one case we have seen on Kalopanax, the stain in the wood was blue.

Further Reading: Piearce and Gibbs, 1981.

VIRUS AND OTHER GRAFT-TRANSMISSIBLE DISEASES

Agents: *Viruses and phytoplasmas*

Damage Type: Disruption of metabolic processes.

Symptoms & Diagnosis: A wide range of non-specific symptoms may be produced (e.g. poor growth, dieback). The commonest distinctive symptoms likely to be encountered on trees in this country are various forms of leaf discoloration: yellow rings, patches, lines (Fig.389, on Syringa), mottles or mosaic patterns (Fig.390, on Populus); or yellow areas centred on the midrib and running along lateral veins; or a mostly yellow leaf with islands and rings of green tissue remaining; or yellow veins (Fig.391, on Sambucus).
Confirmation: A diagnosis cannot be confirmed in the field. Confirmation requires electron microscopy and/or artificial transmission to indicator plants.
Caution: Similar symptoms may be caused by certain herbicides (p.93).

Status, Significance and Host Trees: Poplar mosaic virus (Fig. 390) causes considerable reduction in the growth of infected poplars in the nurs-

Fig. 390

Fig. 391

Fig. 389

SECTION 3

ery but in this country neither it nor any similar leaf disease is known to be harmful to the general health of out-planted amenity trees.

Virtually any broadleaved tree species anywhere may be affected.

Infection & Development: Most viruses multiply and spread in living cells throughout the plant and, depending on the particular virus, are spread from plant to plant by sucking insects, nematodes, mites, fungi, pollen or by direct contact with infected plants. Phytoplasmas inhabit the sieve tubes of infected plants and are spread only by sucking insects. Many of these pathogens, however, (e.g. Poplar mosaic virus) are not known to be transmitted except by grafting or budding, though they are also perpetuated in plants raised from cuttings, and sometimes in seedlings from infected plants.

Remarks: In top fruit growing, as in agriculture and horticulture, uncontrolled viral diseases have the potential to cause substantial reductions in yields; viruses also often prevent the successful propagation by budding and grafting of fruit trees and can be similarly troublesome in the propagation of Flowering cherries and crabs; abroad, several serious diseases of forest trees (e.g. Elm phloem necrosis (= Elm yellows) and Ash yellows in the USA) and Witches' brooms on many species of tree are caused by phytoplasmas; in this country, however, whether these agents play any significant part in the ill health of forest or amenity trees is unknown.

Even though virus particles have often been found in outplanted broadleaved amenity and forest trees (and very rarely in conifers) in this country, the overt symptoms they cause have almost always been nothing more than leaf markings, often inconspicuous and transitory, while any effect they might have on growth has rarely been determined.

Perhaps the virus disease of most potential interest to the arboriculturist in this country is Black line disease of walnut (p.82), since this has killed large, outplanted trees and because similar diseases may be the explanation for other late failures of grafted trees.

Graft-transmissible agents are responsible for a number of the variegated or otherwise prized ornamental trees which can only be propagated vegetatively. (Peace (1962) states that Populus × candicans 'Aurora' is so caused; this is now known not to be the case).

The Mossy willow catkin galls may well be virus-caused (p.184).

Bird cherry and Barley yellows dwarf virus. From time to time, Prunus padus is condemned by some for harbouring, over winter, the aphid vector (Rhopalosiphum padi) of this virus and thus of being in part responsible for the disease. This is a misunderstanding of the aphids' behaviour: the individuals which transmit the virus overwinter on grasses; those which overwinter on P. padus (and P. spinosa) may acquire the virus from infected grasses in spring, but then migrate to barley too late in the year to add significantly to the incidence of the disease.

Further Reading: Cooper, 1979.

VOLUTELLA BLIGHT OF BOX

Fungus: *Pseudonectria rousseliana*
Anamorph: *Volutella buxi*

Damage Type: Bark (and probably sapwood) killing.

Symptoms & Diagnosis: Conspicuous patches of dead, brown, withered foliage appear and other patches of fading, yellowing, dying leaves may be present (Fig.392).

The affected foliage in each patch all arises from one and the same branch. Affected leaves are held close to the twigs instead of at the normal wide angle. Cankers may be present on larger branches.

Confirmation: The bark at the base or part way along affected branches is dead and peels away easily from the underlying wood. The wood in infected branches is characteristically stained grey or black (Fig.393). On the underside of the dead leaves on the plant there may be pink, waxy pustules (spore masses); on fallen, dead leaves may be found minute, greenish or reddish yellow, spherical outgrowths (fruit bodies).

Status: Imperfectly known. The disease has been noted from scattered localities in England and once recently in Scotland. It seems likely that the paucity of records reflects the chronic shortage of plant pathologists working on the problems of ornamental shrubs (box is rarely allowed to reach the stature of a tree) rather than a true picture of the status of the disease.

Significance: Very disfiguring and potentially ruinous in box topiary work, but the lack of records suggests that outbreaks may be sporadic and infections self-limiting.

Host Trees: All records that have been traced are from Buxus sempervirens.

Infection & Development: The fungus overwinters in the fallen, diseased leaves. In spring, spores produced on

Fig. 392

Fig. 393

these and wafted upwards are thought to infect bark wounds, such as those created during hedge clipping; it is also suspected that infection can take place at branch crotches – presumably through the natural stress cracks which form there. Presumably the fungus then grows into the twigs and thence into shoots and leaves as, in suitably moist and warm weather, leaves still attached to infected twigs produce spores from their undersides. These can cause further infections. Infected, dead leaves eventually fall.

SECTION 3

Control: Shortly before growth starts in spring, remove diseased branches and as many fallen leaves from within and beneath the bush as possible. Immediately spray the bush thoroughly with a fungicide based on copper, then again in mid May, mid June and late August. [Take note of current Pesticides Regulations – see p.288]

Remarks: The fungus has been known in England since the mid 19th century. This account is based entirely on the rather sparse published literature.

WATERLOGGING   (with fresh water; for salt water see also p. 93)

Damage type: Malfunctioning and death of roots.

Symptoms and Diagnosis: Leaves are chlorotic or show reddish or purple tinges: leaf margins are browned; petioles may be curved downwards (epinasty). Small, corky outgrowths are present on the underside of leaves or needles (oedema). If the waterlogging is prolonged, or in sensitive species, leaves are small and shoots are stunted, foliage wilted, fruiting poor; defoliation has occurred and perhaps dieback of twigs and even death of whole plants. Odd trees or patches of trees are affected.
Additional indicators: In young trees, the soil around the roots is markedly less heavy than the surrounding clay soil; the stem base of small trees is swollen; the topography or soil type associated with the affected trees is consistent with poor drainage; deposits of silt and debris are present; a water course or drainage ditch is nearby.
Confirmation: Roots are dead, their bark and wood stained bluish or blackish; dead roots and surrounding soil are malodorous; adventitious roots have developed near the soil surface above the dead roots.

Caution: By the time symptoms are reported, the water may have drained away,. any blockage causing the flood may have been cleared and new weed growth may conceal silt deposits. Salt water (p. 93) will complicate these symptoms. Dying alder on sites subject to flooding are likely to be affected by Phytophthora disease (see page 205).

Status: Occasional where ditches have become blocked or drainage systems disrupted during building works, road construction or landscaping, or rivers have flooded in the growing season; more frequent in clay soils where young trees have been planted in holes filled with a lighter, better drained medium to create a sump in which rain water collects.

Significance: Depending on tree species, season of the year, duration of the waterlogging and whether the water was stagnant or flowing, fresh or saline, damage can vary from slight to severe and occur slowly or rapidly, while trees which survive may recover quickly or slowly.

Susceptible and Resistant Trees: Very few species grown in this country can withstand prolonged waterlogging during the growing season. Most conifers are more susceptible than most broadleaves. Particularly resistant are Acer rubrum, Ilex decidua, Larix laricina, Nyssa aquatica and Taxodium distichum (which is also tolerant of saline water); particularly susceptible include Carya glabra, C. ovata, C. tomentosa, Cornus florida, Liriodendron tulipifera, Magnolia grandiflora, Ostrya virginiana and Prunus serotina. A long list of susceptible and intermediate species is given in Sinclair et al. (1993). For the position with alder see above.

Reasons for Damage: Roots starved of oxygen begin to respire anaerobically. This results in an accumulation of toxic products such as ethanol and in an increase in carbon dioxide. Roots then lose their ability to take up water and minerals. An internal water shortage develops and this triggers closure of stomata leading in turn to reduced photosynthesis and translocation of carbohydrates within the tree. Mycorrhizal activity is also reduced and this further reduces mineral uptake. After the soil, water and air content return to normal, drought symptoms may develop because the damaged roots are unable to satisfy the tree's water requirements. Further damage may occur from root infection by Phytophthora, a disease encouraged by wet, especially saturated, soil (see p. 206).

Control: Do not create sumps in clay soils when planting trees but plant into the unamended soil or into mounds of better draining soil. Clear blocked drains promptly. Ensure that soil water drainage is unimpeded around trees during landscaping, building or other site works.

Remarks: Trees can withstand prolonged waterlogging while dormant but waterlogging in the growing season, when the roots' demand for oxygen is high, can be damaging after only a few days and deaths may quickly follow. Flowing water is better aerated and therefore less damaging than stagnant water. Salt water will worsen the damage.

WATERMARK DISEASE OF WILLOW

Bacterium: *Erwinia salicis*

Fig. 394

Damage Type: Blockage of water-conducting vessels; poisoning; weakening and staining of wood.

Symptoms & Diagnosis: During spring or summer, the leaves on some twigs or branches wilt, turn reddish brown and die but remain hanging on the tree (Fig.394). Currently affected parts may arise from leafless twigs and branches partly killed in earlier years.

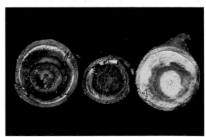

Fig. 395

SECTION 3

Live, dying or dead epicormic recovery shoots may also be present at the base of diseased branches. Occasionally, infected trees show scattered leafless, dead branches but no current leaf symptoms.

Confirmation: If a branch bearing the reddish leaves or the live base of a partly killed branch is cut across, a distinct, watery reddish brown or brownish black stain will be seen in the normally white wood (Fig.395). The stain is likely to extend well back into the otherwise symptomless parts of the tree, sometimes even into the roots. In humid conditions, drops of a sticky, whitish exudate (masses of bacteria) may ooze from the exposed stain and from cracks and crotches on diseased branches (similar to the Fireblight bacterial ooze illustrated in Fig.139).

Status: Occasional in the Cricket bat willow growing areas of Eastern England. Its status elsewhere is uncertain, but confirmed cases are rare.

Significance: A very serious disease of Cricket bat willows as infected wood is brittle and therefore useless for bats. It disfigures but does not often kill ornamentals.

Host Trees: Most records are from the Cricket bat willow (S. alba var. caerulea) and other S. alba varieties and hybrids. Occasional cases have been confirmed on several other Salix species but its full host range and relative susceptibilities are not known.

Infection & Development: The infection biology has not been fully elucidated.

During the growing season, large numbers of bacteria are present on leaves of diseased trees and of healthy trees nearby. It is possible that these, and bacteria which ooze from infected branches, give rise to infections through injuries and through natural wounds such as leaf scars.

There is some evidence to suggest that the bacterium can also spread from infected trees (or stumps, where it can remain viable for at least four years) into neighbouring healthy trees via root contacts.

Strong circumstantial evidence indicates that outbreaks of the disease on previously unaffected sites can originate from the stools used to raise the sets for the planting stock: apparently healthy trees raised from infected but symptomless sets can continue to harbour the bacterium for many years before something, as yet unknown, triggers the development of the full-blown disease.

Control: Legislation exists to protect commercial Cricket bat willow plantations from the disease: see Section 4, p.290.

Therapy: None available.

Prevention: To protect healthy willows nearby, diseased trees should be excavated and burnt.

Alternatively, tops should be burnt, the stumps and roots chipped and the chippings burnt, or disposed of well away from willows. Diseased stumps that cannot be destroyed should be killed with a herbicide, if circumstances allow, as this prevents recurrence of the disease in regrowth and may kill the bacterium in the stump. Species other than willow should be favoured as replacements and willows never planted next to diseased stumps.

If cutting implements used on diseased trees are to be used later on healthy trees they should be cleaned and then sterilized by swilling in or swabbing with a solution of 7 parts methylated spirits to 3 parts of water.

In localities with a history of the disease, large-scale amenity plantings of willows should be avoided.

Remarks: The cause of the disease was first demonstrated in the 1920s, in Essex. Outside England, it is known only from Holland.

Further Reading: Patrick, 1990.

WEEPING CANKER OF CAUCASIAN LIME

Cause: *Probably a bacterium*

Damage Type: Bark killing.

Symptoms & Diagnosis: One of three symptoms may draw attention to this disease: (i) In spring or summer, the foliage of the upper portion or of all of the crown turns yellow and may appear sparse, and/or (ii) Between about August and December, a frothy, thick, milky, sweet liquid exudes, perhaps forcefully and audibly, from the unpruned length of the stem (Fig.396), and/or (iii) Cracks and irregularly shaped, long, narrow, barkless cankers are evident on the unpruned part of the stem (Fig.397).

Confirmation of (i): A patch of dead bark (Fig.398) or a barkless canker encircles the stem just below the point where the branches bearing the yellowing leaves arise. (ii) The exudate flows from dying bark (Fig.396); the wood beneath is stained black or brown. (iii) Cankers often develop very close to the base of branches but rarely involve them (Figs.397 and 398).

Caution: Squirrels may remove bark from stems and branches.

Status: Known from widely scattered localities over most of England.

Significance: The disease causes extensive dieback and lays stems open to infection by wood rotting fungi. It affects a very high proportion of the planting in each case but has been found only on trees between about 10 and 35 years old.

Host Trees: Known only on Tilia × euchlora. Where T. platyphyllos and T. cordata have been growing close to diseased T. × euchlora, they have remained healthy.

Infection & Development: Not fully elucidated. The first indication that bark is infected is the exudation of a

Fig. 396

Fig. 397

SECTION 3

sweet, white liquid very attractive to wasps, butterflies and other insects (see Symptoms above). During winter the discharges cease and the soft, dead bark falls away. Cankers begin to callus over in the following season but at some points on the margins of some cankers bark killing and fluxing resumes. Cankers several feet long but often only a few inches wide commonly form, but some girdle and kill the stem. Vigorous epicormics may develop lower down as a result. Isolations from fresh lesions from several sites have yielded a bacterium resembling Erwinia quercina, the cause of an acorn and shoot blight of evergreen oaks in California which is also characterised by copious fluxes. The bacterium has not yet, however, been shown to cause this disease of limes.

Fig. 398

Control: *Therapy:* None is proven but excision of dying bark immediately fluxing is seen may prevent girdling, though decay may still enter through the wound created.

Prevention: At present, trees older than 35 years seem not to be at risk but, in England, it would be unwise to rely heavily on T. × euchlora in any planting scheme.

Remarks: The disease was first described from Suffolk in 1975. The first report of this disease from outside England was from Sweden in 1993.

WITCHES' BROOMS   Mostly fungi: *Taphrina species*

Damage Type: Abnormal twig proliferation; reduction in flowering on affected parts.

Symptoms & Diagnosis: (Figs.399–404 and 255,256). Conspicuous, abnormally dense, large or small clusters of live or dead twigs (witches' brooms), are evident in an otherwise normal tree crown; the brooms flush earlier than the rest of the tree in spring, bear no or few flowers and lose their leaves prematurely; the leaves are thickened between the veins (Prunus – Fig.404) or crumpled or otherwise slightly malformed and smaller than normal (Betula, Carpinus), and yellowish or reddish; the base of the twigs or branches in the brooms are abruptly thickened (Figs.399,401 arrow);

in Carpinus and Prunus, the brooms are often large and heavy and therefore often pendulous (Figs.399,401,403).
Confirmation: In summer, a whitish bloom (asci) covers the distorted underside of leaves in the brooms. In Prunus, the withering leaves smell of new-mown hay.
Caution: At a distance, clusters of mistletoe (pp. 6 and 103) can be mistaken for witches' brooms.

Status: The Taphrina brooms are common and widespread. Other brooms are occasional.

Significance: An interesting and, to some, an attractive curiosity, though in Flowering cherries, the inhibition in

Fig. 399

Fig. 400

Fig. 401

SECTION 3

flowering may be considered a disadvantage.

Host Trees: Betula pendula (Fig.400) and B. pubescens (caused by T. betulina (= T. turgida)); Carpinus betulus (caused by T. carpini); Prunus (Figs.399,401,404 – caused by T. wiesneri (= T. cerasi)). These brooms are common on Prunus avium and occasional on Japanese cherries. A denser broom on plum and damson is caused by T. insititia. Also on Salix (p.184), Abies and many other species (see Remarks).

Infection & Development: The fungus Taphrina perennates as mycelium in buds and in the bark of twigs. In spring the new shoots and leaves emerge already infected. Windborne ascospores from infected leaves infect further healthy, young leaves. Whether shoot infection is direct or results from

Fig. 402

Fig. 403

mycelial spread via the petioles from infected leaves is not known. See also Remarks below.

Control: If required, to restore full flowering, brooms can be cut out (but prune Prunus species only in summer – see pp.72 and 254).

Remarks: After many years the brooms caused by T. wiesneri on Prunus may become very large (2 or 3 metres long) but are very much less twiggy than those on Betula and so much less conspicuous, especially in winter.

In the past, Taphrina witches' brooms on Betula have sometimes been mistakenly attributed to the mites that are found living in them.

An uncommon Witch's broom on Abies is caused by the rust fungus Melampsorella caryophyllacearum; a number of rare Witches' brooms are caused by other micro-organisms (such as phytoplasmas – see p.263) or invertebrates or have unknown causes (Figs.402, on Pinus sylvestris; 255 and 256 on Salix).

Fig. 404

WOOLLY APHIDS AND ADELGIDS (on Conifers)

Insects: *Adelges spp.*
Pineus spp.
Mindarus spp.

Damage Type: Foliage discoloration and distortion; branch or shoot dieback.

Symptoms & Diagnosis: (Figs. 405–407). White or greyish specks or larger accumulations of waxen wool, often fluffy, are present on needles, shoots, stems or roots and may be more noticeable than the damage itself.

Additional indicators:

On Abies branches and stems: needles distorted – see Silver fir woolly aphid, p.251.

On Abies needles: needles distorted at shoot tip – see Mindarus abietinus under Significance below.

On Abies main stems: crowns deteriorating – see Balsam woolly aphid, p.76.

On Larix needles: needles distorted and discoloured bluish white; Honeydew and Sooty moulds present – see Larch woolly aphids, p.155.

On Picea and Pinus roots: trees showing no sign of ill health – see Stagona pini and Pachypappa tremulae under Significance below.

On Pinus bark of branches and young stems: foliage yellowing; Honeydew and Sooty moulds present – see Pineus pini under Significance below.

On Pinus strobus main stems: trees apparently healthy – see Pineus strobi under Significance below.

On Pseudotsuga needles: needles distorted; Honeydew and Sooty moulds present (p.144) – see Adelges cooleyi under Significance below.

Significance and Hosts: The more damaging species are given separate entries as indicated above.

Less damaging species:

On Abies needles: Mindarus abietinus can considerably disfigure needles at the shoot tip, causing them to twist. The damaged needles are retained (Fig.405).

On Picea and Pinus roots: Stagona pini and Pachypappa tremulae produce copious quantities of waxy wool on the roots of Pinus and Picea respectively. These are only a problem in nurseries and on container-grown plants (Fig.406). They do not cause damage to established amenity trees.

On Pinus bark: Pineus pini can cause yellowing and loss of needles leading to shoot dieback on drought-stressed P. sylvestris and P. contorta (Fig.407).

On Pinus strobus main stems: Pineus strobi causes no obvious damage despite its occasional great abundance.

On Pseudotsuga needles: Adelges cooleyi (the Douglas fir woolly aphid) is a damaging pest in young forest plantations, stunting growth and causing yellow blotches on infested needles, but is not a serious problem on established amenity trees.

Life Cycle: Adelgids have complex and varied life cycles. They may be heteroecious, i.e. with an annual alternation of generations (not always obligatory) between Picea and another host (e.g. A. cooleyi on Pseudotsuga), or autoecious with the whole life cycle completed on one host (e.g. Pineus spp. on Pinus and A. piceae on Abies). They can be anholocyclic, reproducing throughout the growing season by parthenogenesis (no sexual forms occur), or holocyclic with the parthenogenetic cycle being interrupted by an annual sexual phase when males and females are produced.

SECTION 3

Fig. 405

Fig. 406

Control: Specific recommendations for the more damaging species are given under their separate entries, as indicated under Symptoms above. Otherwise, when necessary, use a persistent contact insecticide, approved for this purpose, on a mild and calm day between early November and the end of February. Apply the insecticide at high volume (i.e. spray to run-off) to ensure contact with adelgids overwintering close to terminal buds. [Take note of current Pesticides Regulations – see p.288]

Remarks: Aphids and adelgids, together with a third group, the phylloxerids, are considered to be members of two closely related superfamilies, the Aphidoidea and the Adelgoidea, although there is still disagreement about the precise relationships of these groups (which are treated in a different way by some taxonomists). Most aphids, however, can be recognised by the presence of paired siphunculi (cornicles) on the abdomen, although these

Fig. 407

are absent in a few genera. Adelgids and phylloxerids do not have siphunculi and their antennae are much shorter than those of aphids. These two groups also hold their wings roof-like (adelgids) or horizontally (phylloxerids) over the body. In contrast the wings of true aphids are held together vertically above the body so that individuals may be picked up by holding the wings between finger and thumb.

Aphids differ biologically from both other groups in that the parthenogenetic females are viviparous (give birth to live young) whereas those of adelgids and phylloxerids are oviparous (lay eggs). Adelgids only feed on conifers while phylloxerids are restricted to broadleaved trees. As a group the true aphids feed on the sap of a wide variety of broadleaved and coniferous trees.

Further Reading: Carter (1971), Bevan (1987), Carter & Winter (1998).

WOOLLY APHIDS AND SCALE INSECTS
(on Broadleaves)

Fig. 408

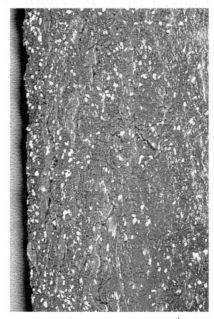

Fig. 409

SECTION **3**

Damage Type: Unsightly wool on bark. See also Significance below.

Symptoms & Diagnosis: Waxy wool, usually white, in small tufts and spots or larger patches, is present on the branches, trunk, exposed roots or root collar and is sometimes associated with dimpled, pitted, fissured or roughened bark or pruning wounds (Figs.408,409). *Confirmation:* Mingled with the wool will be found either aphids or scale insects. Scales: a hardened tortoise-like shell up to 7–8mm long, surrounded by wax (see e.g. Fig.187). Aphids and felt-ed scales: very small soft bodied insects (most often yellowish to purplish-brown) hidden amongst the wax from July onwards. Woolly aphids on Malus and other rosaceous hosts occur on current shoots from July onwards. Honeydew and Sooty moulds are often present (p.144).

Caution: Waxy wool on the leaves is also caused by psyllids but is then associated with galls (see p.86).

Status: Occur throughout Britain wherever the host trees are grown.

Significance: Most species cause little harm. Exceptions are the Felted beech scale on Fagus (p.126), which may lead to the development of Beech bark disease (p.77), and the woolly aphid Eriosoma lanigerum (also misleadingly known as American blight), which is often a nuisance on ornamental trees (see hosts below). Splitting of the bark where aphid colonies persist on Malus can lead to apple canker. Damage to trees, however, is usually less serious than the amounts of waxy wool suggest (Fig.408).

Host Trees:

Acer, Aesculus, Tilia and other species:
Pulvinaria regalis (p.149)

Crataegus, Malus (the commonest host), Pyrus, Sorbus and other rosaceous trees:
Eriosoma lanigerum (a woolly aphid) (Fig.408)

Fagus:
Cryptococcus fagisuga (p.120)
Phyllaphis fagi (p.80)

Fraxinus:
Pseudochermes fraxini (a felted scale – Fig.409)

Ulmus:
Gossyparia ulmi (a felted scale)

Life Cycles: For scale insects see pp.248.

The woolly aphid E. lanigerum overwinters in bark cracks or under loose bark and is inconspicuous at this stage. Wool is produced from March or April and the breeding colonies develop from May onwards throughout the summer, reproducing by parthenogenesis (females only) and giving birth to live young. Some winged aphids appear in July and may fly to other trees. Breeding slows down in autumn. Some eggs are laid in September but these are usually infertile.

Control: *Scale insects:* Not necessary except in special circumstances for C. fagisuga (see p.120) and P. regalis (see p.149). *Woolly aphids:* E. lanigerum may be controlled during the dormant season with tar oil winter wash. In spring and early summer treat with a suitable aphicide. Avoid spraying when trees are in blossom because of danger to bees. [Take note of current Pesticides Regulations – see p.288] For local control, individual colonies can be dabbed with a stiff paint brush in May or June: the bristles will kill the aphids. This treatment is more effective if the brush is dipped in paraffin or methylated spirit but do not then use on soft green stems, shoots or leaves.

Remarks: Eupulvinaria hydrangeae, a species found recently to occur in Britain, is similar to P. regalis but the waxy wool masses occur on the leaves and twigs, rather than on the trunks. In Britain E. hydrangeae has only been found on Hydrangea in the London area, but in Europe, where it was first found about 1965 having originated possibly from Japan, it is recorded from various trees including Acer, Cornus, Crataegus, Platanus and Tilia.

YEW DIEBACK AND CANKER

Cause: *Probably a fungus*

Damage Type: Bark killing.

Symptoms & Diagnosis: In clipped hedges, patches of yellow or brown foliage or leafless twigs are evident (Fig.410). On close examination, each patch of foliage is found to belong to a single branch. On unclipped trees, the foliage on one or two small or large branches is yellow, brown or dead (Fig.411).

Confirmation: Cutting reveals that at some point along an affected branch, below affected foliage, there is a girdling patch of dead bark; or the base of the branch is surrounded by a bark canker (Fig.412).

Additional indicators: Sunken, elliptical bark cankers centred on dead twigs are present on some live branches.

Caution: Scattered branches in clipped yew hedges are occasionally girdled and killed by the gnawing of Bank voles (p.176). Shoots may be killed by the Fruit tree tortrix (p.131).

Status: Probably a common but only occasionally a conspicuous disease. It has been noted from scattered localities from Sussex and Wiltshire northwards to Derbyshire and Sheffield.

Significance: The disease can badly disfigure clipped hedges and topiary if neglected and allowed to spread back into main branches or the stem. It spreads slowly, however, so is readily controlled by surgery and, if caught in time, regrowth will fill in the gaps.

Host Trees: All records to date have been from Taxus baccata; a few cases of a similar, perhaps the same, unexplained disease have been seen on Cedrus.

Infection & Development: Not yet fully elucidated, but the following seems to be the likely sequence of events (the pathogen is currently thought to be a fungus): at some time during the year, spores are produced from infected bark. During the tree's dormant season, the pathogen grows from the site of infection (observations indicate that both pruning wounds and young tissues such as buds, needles or shoots are susceptible) into the surrounding bark, killing it. As the fungus spreads, a bark canker is formed which may eventually girdle the member,

SECTION **3**

Fig. 410

causing the distal part to die; or a long canker may develop on one side of a branch, girdling and killing any side branches in its path. The fungus may also spread back steadily and evenly along the twig or branch causing a true dieback.

Control: *Therapy:* Cut dead and dying branches off below the lowest extent of any dead bark, preferably back to a side branch or shoot.

Prevention: None known. The apparent infrequency with which pruning wounds are infected does not warrant the routine use of a wound protectant.

Remarks: First noted in the 1960s, in England. Attempts to infect yew trees artificially with various fungi isolated from the bark lesions have been unsuccessful. No organism has been found consistently associated with the damage.

Fig. 411

Fig. 412

YEW GALL MIDGE (ARTICHOKE GALL)

Insect: *Taxomyia taxi*

Damage Type: Formation of dense needle clusters.

Symptoms & Diagnosis: In summer the new shoots develop into a tight cluster of terminal leaves, the Artichoke gall (Fig.413 arrow). These occur most often on shoots which develop from terminal buds but shoots from lateral buds are also affected. The galls persist and become larger during the second year, eventually up to 35mm diameter and 25mm high. The outer leaves of the galls are dark green, the inner leaves yellowish or white. Vacated galls turn brown and may remain on the tree for another two years.

Confirmation: Cut green galls open to reveal orange larvae within (Some galls will be empty or will contain the whitish larva of a wasp that has parasitized the gall midge).

Status: Widespread, probably wherever Taxus grows in Britain.

Significance: Abundant on some trees, rarely seen on others. Disfiguring on

Fig. 413

amenity trees. Continuous attacks can stunt young trees, otherwise of little importance.

Host Trees: Taxus spp; also recorded on Cephalotaxus.

Life Cycle: The adult midges occur in May and June when the eggs are laid singly in the buds. The larvae develop inside the galls for one or two years.

Control: If necessary handpick infested galls on small trees. No control is necessary on larger trees.

SECTION **3**

EXOTIC PESTS AND DISEASES

Although there is no definitive list of undesirable aliens, organisms tolerable in their native countries have often become noxious pests when exported. The following, well known to be damaging to trees abroad, have the undoubted potential to cause serious damage to some of our most valuable amenity and timber trees should they reach our shores and become established here. Only the salient features are briefly described here; more detailed information is obtainable from the Forestry Commission's research stations. Any suspected cases should be reported to your local MAFF Plant Health Inspector (address in the telephone book) or to the Forestry Commission (addresses on p. 306).

CANKER STAIN OF PLANE

Fungus: *Ceratocystis fimbriata* f.sp. *platani*
Host Trees: *Platanus*
Now present in Continental Europe

Symptoms: Late in the disease's development, large or small patches of yellow, dwarfed leaves appear in the crown. With or without these foliage symptoms, elliptical or long, thin patches of dead bark, small or very large, are present on trunks or limbs in association with pruning or other bark wounds. Behind and beyond the areas of dead bark the wood is stained reddish-brown or bluish-black and is often infected by wood-rotting fungi. The crown deteriorates over a number of years before the tree dies.

CHESTNUT BLIGHT

Fungus: *Cryphonectria parasitica*
(= *Endothia parasitica*)
Host Trees: *Castanea sativa*
Now present in Continental Europe

Symptoms: The foliage on one or more branches wilts and dies as the result of girdling cankers lower down on them or on the trunk. On young stems, the contrast between green healthy bark and orange diseased bark can be very striking. If the dead bark is removed conspicuous fans of buff-coloured mycelium can be seen in the cambial region. A few weeks after the bark has died it becomes studded with innumerable pimples a millimetre or two across from which, in wet weather, yellowish or buff-coloured spore tendrils ooze.

| OAK WILT | Fungus: *Ceratocystis fagacearum*
Host Trees: *Quercus*
Now present in North America |

Symptoms: In many ways, much like Dutch elm disease (see p. xx): leaves over part or much of the crown wilt, discolour and fall; twigs, branches and whole trees die. The whole tree may show symptoms within a few days or a few weeks or may die back over several years, during which time adventitious recovery shoots develop on limbs and trunk and die in turn.

In red oaks (infrequently in white oaks like our native Q. petraea and Q. robur), a few months after defoliation, mats of grey mycelium with a raised central area may develop beneath dead stem bark, forcing it to split open. The elliptical mats are up to 10 cm wide and 20 cm long and have a rich fruity smell. Adjacent trees may become infected via root grafts to give rise to enlarging groups or rows of diseased trees.

| ASIAN LONGHORN BEETLE | Insect: *Anoplophora glabripennis*
Host Trees: Wide range of broadleaved trees including *Acer, Aesculus, Alnus, Malus, Populus, Prunus, Salix, Ulmus*
Now present in China, elsewhere in Asia and in North America. Intercepted several times on imports to this country |

Symptoms: Death of branches, tree tops or whole trees. Round, adult beetle emergence holes, 9-11 mm diameter, appear in May-August (-October) in infested bark. Masses of wood shavings will be present around and on the ground below fresh emergence holes. Removal of the loose bark will reveal larval galleries leading to holes up to 10 mm in diameter penetrating into the wood. The adult beetle has a striking appearance.

| EIGHT-TOOTHED SPRUCE BARK BEETLE | Insect: *Ips typographus*
Host Trees: *Picea*
Now present in Europe and Asia |

Symptoms: Needles turn red-brown and then fall. Bark on main stem is peppered with round, adult beetle emergence holes 1.5 - 2.0 mm diameter. Beneath the dead bark, gallery systems have two or more vertical galleries leading up or down from a central chamber with a mass of side tunnels leading away across the cambium. A fungus associated with the insect contributes to tree death.

SECTION **3**

WHITE PINE WEEVIL	Insect: *Pissodes strobi* Host Trees: *Picea, Pinus* Now present in North America

Symptoms: In July, needles turn red on leading shoots. These shoots wilt, die and later fall off. Damage is caused from May to July by white larvae boring upwards beneath the bark.

SIBERIAN FIR WOOLLY APHID	Insect: *Aphrastasia pectinata* Host Trees: *Abies* Now present in Siberia and east to Poland, Finland, Sweden and Norway

Symptoms: Needles of all ages discolour, brown and fall; shoots die; prolonged attacks may kill trees. Infested live needles show yellow spots on the upper surface with corresponding white, woolly tufts beneath concealing the aphid.

EUCALYPTUS BORER	Insect: *Phoracantha semipunctata* Host Trees: *Eucalyptus* Native to Australia. Now also present in Spain, Portugal and around the Mediterranean

Symptoms: Branches or whole trees killed by larvae burrowing in the sapwood.

SECTION 4

CONTROL OF AMENITY TREE PESTS, DISEASES AND DISORDERS

CONTROL OF AMENITY TREE PESTS, DISEASES AND DISORDERS

Although this handbook is chiefly concerned with diagnosis and, therefore, the control of existing problems (therapy), the reader will also wish to know how to prevent the problem recurring on the same or other trees or on trees yet to be planted (prevention). These two aspects of control are therefore dealt with separately here.

The special problems of preventing, detecting and treating decay are discussed separately in Section 5.

A. THERAPY

Once the diagnosis has been made, it will be necessary to decide what action, if any, should be taken. An entry under 'control' is therefore to be found under each problem described in Section 3. However, for reasons given below, even some available control measures need not or cannot be implemented. As the decision will depend on circumstances, the suggestions given in Section 3 should be regarded as what could be done; what should be done is discussed below.

Should control measures be instituted?

In order to decide whether to take any action, the following seven criteria should be considered. If they are taken in the given order, it may be unnecessary in any particular case to consider every one of the seven.

1. Is control desirable?

If untreated, would the damage be so great or of such a nature that the tree would no longer fulfil its purpose? For example, the perennial loss of flowering in a purple leaved plum from bird damage may well impair the tree's primary role as an ornamental but its handsome foliage may nevertheless be considered justification enough for its retention.

Tree pests and diseases and those disorders caused by natural phenomena are as much a part of the natural world as the trees themselves and intrinsically as worthy of preservation and study as the trees are. Provided that they do not too greatly interfere with the purpose for which we are growing our trees, or render them dangerous, or destroy them, control is unnecessary. Furthermore, ancient or decrepit trees are home to many interesting non-pathogenic and often rare fungi, insects and other organisms which are themselves worthy of conservation.

Taking this view, 'problems' can diminish, become acceptable, even welcome, in the same way that to a nature lover, a 'weed' may become a 'wild flower'.

2. Have effective therapeutic measures been devised?

For many amenity tree problems, no cure or palliative has been devised, although many problems can be prevented or avoided (see 'Prevention' below). For others, measures worked out for similar problems may be effective (e.g. all powdery mildews are likely to be susceptible to the same fungicides).

3. Will control measures applied now be effective?

This depends on the stage of development of the condition as well as on the nature of the treatment. The timing of fungicidal and insecticidal treatments is often crucial to success; where excision is an option, timing is often less critical.

4. Is control necessary?

Often it is not. It depends largely on the severity of the effect of the disorder on the purpose for which the tree is being grown (though where the problem is one of wood decay or other structural weakening, public safety must also be considered). If the effect is only slight or temporary then control is not strictly necessary. There is no merit in taking action for the sake of 'doing something' or just because it can be done.

The diagnosis may indicate that action will be necessary but can be deferred: for example, crab apples defoliated by scab each year can be replaced gradually to avoid denuding the whole street of trees at once.

5. Is control practicable?

It may be impossible to excavate an infected stump, for example, without damaging adjacent underground pipes or cables. Funds or suitably skilled staff may not be available to implement control measures.

6. Is control permissible?

Strict environmental pollution and public health regulations may prevent the application of effective pesticides even though the tree problem may warrant such control measures (See 'Pesticide Regulations' below).

Other control measures may also be disallowed: for example, the excavation of stumps to control a root disease may be prohibited on an archaeological site.

7. Is control justifiable?

This will usually depend on whether the expected benefits would outweigh the cost, effort and disruption involved. The result of doing nothing must be taken into account. How the value of the tree is calculated will greatly influence the decision (Anon. 1990).

Curative strategies

Success with a curative treatment usually requires action early in the development of the problem; often by the time a diagnosis has been made it is too late.

a) Biotic problems

i) **Pesticide application**. Few disease or pest problems of trees can be 'cured' in the sense that the pathogen is destroyed so that the tissues can resume their normal function. One exception are the powdery mildews which initially do not kill the host tissue and, being largely superficial, can be destroyed with fungicides. Most 'therapeutic' pesticides, having killed or inhibited the pathogens, allow the plant to function with the remaining undamaged tissues or to produce new, healthy tissue. Often the pesticide also confers a degree of protection against re-invasion.

Most therapeutic pesticides are sprayed or painted directly onto the causal organism but a few pests or diseases can be controlled by injecting the pesticide into the transpiration stream (e.g. Dutch elm disease p.116).

ii) **Surgery**. Often effective but sometimes disfiguring is the simple expedient of removing the affected tissues although the resultant wounds can themselves become diseased. Reduce this risk by following good pruning practice (Anon. 1989) and by taking, when necessary, precautions against Silver leaf (p.252) and Bacterial canker of cherry (p.70).

A tree stripped of bark and girdled by an animal can sometimes be saved by grafting live bark across the gap, or by similar means (see p.181).

b) Abiotic problems

Except in the case of drought, the transitory nature of most damaging meteorological phenomena gives no opportunity for therapy. Treatments for the bulk of the remaining abiotic problems fall into 3 main categories:

  i) Improvement in soil aeration (e.g. Compaction, p.181; gas, p.96; waterlogging p.267).

  ii) Removal, dilution or deactivation of harmful chemicals (e.g. herbicides, salt, p.98).

  iii) Application of fertilizers to correct nutrient deficiencies (seek specialist advice).

B. PREVENTION

Preventative strategies

This section outlines the chief strategies available for the prevention of diseases, disorders and pest problems. Which to choose in any particular case may be self-evident (careful adherence to instructions for the use of a herbicide; fencing out animals). In other cases, appropriate measures are either indicated under the 'control' entries in Section 3 or should become clear once the nature of the problem is understood.

1. Using resistant and tolerant trees

The variation among trees in their susceptibility to various disorders should be taken into account when the initial species selection for planting is made: salt-tolerant species chosen for the seaside, calcifuge species avoided on limestone soils and so on.

Such means of avoiding problems are also applicable to various infectious diseases and other types of damage. For example a Honey fungus infected site could safely be planted with a resistant species like yew.

2. Removing or destroying the causal agent

Causes of damage which do not too readily recolonize or recur can be removed: weeds can be killed or a waterlogged site can be drained; the danger of Honey fungus infection can be overcome by excavating all infected material from the site.

The severity of outbreaks of some pests and diseases can be reduced if the over-wintering stages can be destroyed before these give rise to primary infestations in the spring. Examples are the destruction of twigs killed by the cherry Blossom wilt fungus (p.85) and the pruning out of Brown-tail moth 'nests' before the overwintering larvae in them cause serious defoliation in the spring.

SECTION 4

In this country, the danger of transmitting infection by means of contaminated cutting implements need be taken into account only in dealing with Fireblight (p.124) and Watermark disease (p.267), when the causal bacteria can be destroyed by sterilizing the tools. Tools used on the Continent should be thoroughly cleaned before being used in this country again. In addition, if they have been used on plane trees (*Platanus*) they should be sterilized as described under fireblight (p.126) as a precaution against introducing the Canker stain disease (see page 280).

3. Excluding pathogens and pests

The spread of fungi which grow along roots or through the ground can be interrupted if a barrier is inserted in their path (see Honey fungus, p.148).

The risk of contaminating a site with the soil-dwelling fungal root pathogens *Verticillium* and *Phytophthora* (pp.261 and 206) can largely be guarded against by purchasing stock from healthy nurseries.

A few diseases which are spread by insects can be restricted by limiting the numbers of these vectors: the spread of Dutch elm disease can be considerably influenced if trees killed by the disease are destroyed before beetles carrying the fungus can emerge from them (p.118).

Some mammals can be fenced out or otherwise excluded (see Pepper,1988).

4. Applying protective pesticides

The common agricultural and horticultural practice of coating susceptible tissues with a protective pesticide is rarely practicable, justifiable or permissible on outplanted amenity trees. An exception is the treatment of pruning wounds on some species against, for example, Silver leaf (p.254). (See also Preventing Decay, p.296).

5. Rendering conditions unsuitable for damage development

Such factors as water stress, radical site changes, environmental pollution and poor nutrition can themselves damage the tree. Moreover, trees suffering from such influences are more liable to attack by certain, normally weak, pathogens (e.g. *Cytospora chrysosperma* p.242; *Eotetranychus tiliarium* p.172; *Cryptostroma corticale* p.257). Avoidance of the former will,therefore, also reduce the risk of the latter.

However, indiscriminate attempts to promote a tree's health by 'improving' nutrition can just as well exacerbate or lead to a problem as ameliorate it. Unless there are sound reasons for doing so, do not apply fertilizers or manures.

Other examples of this strategy are weeding to prevent vole damage (Davies & Pepper, 1990), keeping foliage dry to prevent Peach leaf curl (p.197), and employing herbicides or mulches instead of potentially damaging mowing machines to weed around specimen trees (Patch & Denyer 1992).

C. PESTICIDES REGULATIONS

British law controlling the sale, storage, use and disposal of 'pesticides' (the definition includes insecticides, fungicides, herbicides and the like) was radically changed and strengthened with the introduction of the Control of Pesticides Regulations 1986 under the Food and Environment Protection Act of 1985. As a result, and as the situation changes with the introduction of new products and phasing out of others, a number of pesticides recommended in the past – and some in this handbook – are either no longer marketed or, even if still available, may no longer be used for every purpose for which they are suitable. A further complication is that some products now available to professional gardeners and arboriculturists may not be used by non–professionals (amateur gardeners, for example) – although professionals may use 'amateur' products.

The law relating to the use of pesticides on and around amenity trees is briefly as follows (for nurseries, orchards and Christmas tree crops it is different):

1. Before any pesticide can be sold, stored, supplied, used or advertised, it must be Approved (by an appropriate Minister).

2. Most uses to which an 'Approved Product' may be put, the situations in which it may be used, the means by which it may be applied, safety precautions to be observed, and other constraints on its use must be stated on the product label or on any accompanying leaflet. It is illegal to contravene these conditions of use except that:

3. From time to time, Approval is granted for existing Approved products to be used in situations or for purposes not stated on the label – so called 'Off label' Approvals.

Such Approvals are usually granted at the request of a grower for a minor use the demand for which is considered by the manufacturers to be too small to justify the expense of obtaining full Approval.

Off label uses are still governed by the safety regulations which apply to full Approvals but the material is used at the user's own risk (i.e. the manufacturer cannot be held responsible for consequent damage to the plant or failure of the treatment).

Anyone who wishes to use a product under the 'Off label' Approval must first obtain and read a copy of the Notice of Approval, which is available from local Agricultural Departments and Farmers' Union offices.

4. Any user of an Approved Professional product who was born later than 31 December 1964 must hold a Certificate of Competence (issued by the National Proficiency Test Council) or be overseen directly by a Certificate Holder.

5. Professional users of pesticides must also comply with the Control of Substances Hazardous to Health Regulations 1988 (COSHH).

Further reading

- The most generally useful publication for the practitioner (although products Approved for amateur use are not included) is the *UK Pesticide Guide* for the current year published jointly by the British Crop Protection Council and CAB International. This lists currently Approved Professional products (On label and Off label) with their active ingredients and approved uses together with general guidance on their safe and effective use and a summary of the relevant legislation. Obtainable from BCPC Publications Sales, Bear Farm, Binfield, Bracknell, Berkshire RG42 5QE.

- Approved Amateur (and Professional) products (but NOT their recommended uses) are listed in *Pesticides* for the current year published by The Stationery Office. This includes a detailed account of the legislation and the Approvals process and also lists pesticides approved for use as wood preservatives and other non-crop purposes. Obtainable from The Stationery Office, PO Box 29, Norwich, Norfolk NR3 1GN.

- A list of currently Approved Amateur Products together with a list of their permitted uses is available on request to members of the Royal Horticultural Society from Wisley Gardens.

SECTION 4

- The Code of Practice for the Safe Use of Pesticides on Farms and Holdings (PB3528) is obtainable free from MAFF Publications, ADMAIL 6000, London SW1A 2XX. It covers the requirements of both the Control of Pesticides Regulations 1986 and the Control of Substances Harmful to Health Regulations 1988.

D. LEGISLATION CONCERNING OUTPLANTED AMENITY TREES

Our plant health legislation is intended to prevent the introduction of pests and diseases now absent from this country and to inhibit the spread or limit the impact of certain pests and diseases already here .

As circumstances change so do the regulations. Some significant changes concerning trees took place in 1993 with the advent of the European Community Single Market. The current situation can be ascertained from the addresses given at (d), (e), (f) and (g) in Appendix 2, p.306.

The background to the 1993 legislation is usefully summarized in Phillips and Burdekin, 1992.

At the time this handbook was being prepared, there were regulations of some interest to arboriculturists as follows:

1. Dutch elm disease (p.116)

Local authorities in parts of Sussex and in parts of the country north of a line from the Mersey to the Wash were empowered to order the destruction of elms infected by Dutch elm disease.

2. Great spruce bark beetle (p.138)

There were restrictions on the movement of spruce wood grown in designated areas of Wales and the western half of England.

There were restrictions on the movement of all conifer wood with bark in designated areas of Wales and the adjoining English counties.

3. Watermark disease of willow (p.267)

Local authorities in some English counties were empowered to order the destruction of willows infected with Watermark disease.

4. Fireblight (p.124)

In 1993, Fireblight ceased to be a notifiable disease except where it occurred on premises registered for trade under the Plant Health (Great Britain) Order 1993 or in associated 'buffer zones'.

Notifiable cases of the disease should be reported to plant health inspectors of the Ministry of Agriculture, Fisheries and Food or those of the Departments of Agriculture of the Scottish or Welsh Offices (for addresses see Appendix 2, p306).

5. Exotic pests and diseases (p.280)

A number of exotic pests and diseases are deemed likely to pose particular threats to trees in this country should they ever find their way here. These are briefly described on pp. 280-282. Suspected cases must be reported either to MAFF or to the Forestry Commission. For addresses see Appendix 2, p.306.

DECAY AND SAFETY

This section summarizes for the diagnostician the signs and implications of decay and the procedure for determining whether a decayed tree should be regarded as dangerous. It also outlines the various practical steps available for dealing with decayed and dangerous trees, discusses the prevention of decay and gives some useful references. The whole subject is dealt with in comprehensive detail by Lonsdale (1999).

LEGAL LIABILITY

'In law, where a tree shows direct external evidence of decay, its owner is normally liable for any damage it causes if it breaks or falls. The situation regarding indirect evidence of decay is less certain . . . The courts have accepted the principle that people with responsibility for trees, whether owners, tenants or agents, should inspect their trees at regular intervals. Where, on such an inspection, symptoms of ill health or unusual growth are observed, expert advice should be sought. Failure to obtain or act upon such advice could lead to allegations of negligence.' (Young, revised Lonsdale, 1984, p.3.).

INSPECTING TREES FOR SIGNS OF DECAY

'. . . inspections should be made annually, preferably on clear days in September or early October when ephemeral sporophores are usually well developed, and when the crown symptoms of extensive root decay are likely to be so far advanced that they contrast sharply with the dense green foliage of healthy crowns. Binoculars are essential for the adequate inspection of the upper part of large trees and even at relatively close quarters much useful information that would otherwise be missed can often be seen with their aid.' (Ibid. p.9.)

SIGNS OF DECAY

- **Fruit bodies of decay fungi** may appear on stem, limbs or root buttresses or on the ground above decayed roots. The commonest species are illustrated as Figs. 316-352 and described on pp.228-234. Illustrated descriptions and information on the significance of all of them can be found in Lonsdale 1999. Many root-rotting fungi fruit on the stem at ground level and so are liable to be concealed by vegetation, or resemble the bark from which they are growing, or are otherwise easily overlooked. Some fruit bodies are perennial, increasing in size for years, but others appear only intermittently and die quite quickly. The latter may persist for a while as rotting, fleshy remains or dried, denatured specimens, perhaps fallen from the tree, quite unlike the fresh specimens depicted in books but nevertheless recognisable as fungi and sometimes as characteristic of the species as fresh specimens (see e.g. *Laetiporus sulphureus* **Figs.345** and 346).
- **Wood exposed by injuries** may be evidently decayed, but decay may also be concealed behind the apparently sound exterior of old, open pruning wounds. These should be probed (a sharp blow with chisel and mallet should suffice).
- If the fungus is killing living tissues (often the case with root infecting fungi), the tree is likely in time to show signs of ill health (as described in Section 1, Table 3, p.14) as more and more conducting and storage elements in the wood are rendered useless.

SECTION **5**

294 DIAGNOSIS OF ILL-HEALTH IN TREES

- Rotted roots may snap and allow the **root plate to heave** perceptibly in the wind.
- Rotting wood may be concealed beneath **dead bark**. Dead bark may show up as a darkened, sunken, cracked or roughened area, or may be peppered with insect exit holes, or covered in fungal fruit bodies. If tapped with a mallet, it will sound hollow if it has lifted away from the wood beneath. It may, though, look no different from living bark.
- Some decay fungi which kill living bark (e.g. Honey fungus) sometimes induce **gum, resin or a watery fluid** to issue from the dying bark.
- Decay sometimes develops in **tight (acute angled) forks** when bark has been compressed between the members, or when cracks have exposed the wood to infection in the structurally weak joints.
- An **abruptly bent limb** may indicate the earlier loss of a branch end and therefore decay in the old wound.
- A decayed limb or stem may show **horizontal or vertical cracks** before breaking completely.

But a decayed tree, especially one infected by a heart-rot which has entered via the roots, may appear quite normal, showing none of these signs. No means of recognizing such a tree is at present available except the method of probing it as described under Failure-risk Assessment below, or by examining roots and root collar as described on p.8 under 'Examining the Seat of Damage'.

IS THE TREE DANGEROUS?

To decide this point it is necessary to assess firstly the chance that it will fail (failure-risk assessment) and secondly the amount of damage that might result should it fail (consequential damage assessment): a decayed tree is not necessarily dangerous. Objective methods of assessing the failure risk are now available to the specialist (Mattheck & Breloer (1994); Lonsdale (1999)).

FAILURE-RISK ASSESSMENT

Structurally weak points, such as tight forks, must also be taken into account when assessing a tree's safety, and even perfectly formed trees will break under excessive loads, but these factors fall outside the field of 'ill-health' covered by this Handbook.

To what degree the integrity of a decayed tree is compromised depends on the extent of the decay, its severity, its location in the tree, the amount of sound wood which remains to support the decayed part, and the configuration of the tree in relation to the decayed part.

Much can be deduced from a careful visual examination of the tree, but if a more precise assessment of the extent and severity of decay is required, the use of some kind of probe is necessary.

A simple metal rod can be used to probe the depth of severe, exposed decay, and a rough assessment of severe aerial decay beneath sound wood can be obtained with a brace and long twist-bit. Much better, though laborious, is the Pressler increment borer which extracts intact pencils of wood (Phipps 1985). These enable decayed wood to be observed directly and allow the strength of the remaining sound wood to be tested by means of a hand-held device, the Fractometer. They can also be used for growth measurements and laboratory examination.

A disadvantage of all types of auger is the relatively large hole they make which can provide a passage through which the fungus can spread into wood which may otherwise have remained uninfected. The Shigometer (and similar instruments) causes very small wounds – but considerable practice is required before the electrical resistance readings it gives can be reliably interpreted.

The portable compression strength meter (Barratt *et al* 1987) which differentiates between sound and decayed wood by measuring the wood's physical resistance to penetration, is portable, cheap and simple, but laborious to use and creates a relatively large hole. Electrically powered probes which use the same principle, such as the DDD 2000 and the Resistograph, do not provide a core and are much more expensive. However the hole they drill is very narrow, they record data automatically, and their use requires little effort.

Stress wave timers such as the Metriguard require no holes to be drilled. These measure the length of time it takes for the sound of a hammer blow on one side of the tree to pass through the stem to the other side. Sound waves travel more slowly through most decayed wood than through healthy wood. However, until at a very advanced stage, some decay types (e.g. that caused by Ustulina deusta) do not slow sound waves down and so cannot be detected in this way.

Because of the usual inaccessibility of any but the superficial roots, the presence, extent and severity of root decay usually has to be deduced from the signs of decay listed above: digging can be revealing but it is often extremely difficult, impracticable or impossible to assess root decay by this or any other direct method.

The identity of both the fungus and the tree should be taken into account as some fungi are far more aggressive than others and some trees much more rapidly decayed than others. (Burdekin 1979; Greig 1981).

More detailed information on decay detection, tree safety evaluation and allied topics is given in Lonsdale (1999) and Mattheck and Breloer (1994).

CONSEQUENTIAL DAMAGE ASSESSMENT

Having judged the tree's chances of failure it is necessary to judge how much damage might be caused should it fail. This depends on the size of the tree, what it might hit, and in particular whether it might injure people.

DANGER ASSESSMENT

Taking both the failure-risk and consequential damage assessments into account, the degree of danger posed by the tree (its hazard rating) can be judged – for example, a very badly decayed tree in a remote area may pose less danger than a slightly decayed tree in a busy children's playground.

Thought-provoking discussions on this subject are given by Helliwell (1990), Mattheck and Breloer (1994) and Lonsdale (1999).

HAZARD REDUCTION

General considerations

The outcome of an infection depends not only on the activity of the fungus but also on the behaviour of the tree: in response to wounding and infection the tree musters physical and chemical defences which limit the ultimate extent of fungal decay. Thus, decay which has entered through aerial wounds is usually confined to the wood present at the time of wounding, and the spread of decay will slow down or cease once the wound has occluded. Furthermore, the tree is laying down new wood each year which, if the decay process has ceased, slowly begins to compensate for the strength lost to decay or, if decay is continuing, may counterbalance the effect of the decay process.

SECTION 5

Remedial Measures

- Dangerous trees can be felled; dangerous branches can be removed, though new wounds created in so doing are themselves liable to decay in time.
- Where a decayed tree is judged to require remedial action short of felling, cable-bracing, propping, crown thinning or crown reduction may be sufficient to lessen the risk of failure appreciably, at least temporarily.
- As an alternative or in addition to pruning or removing the trees, consideration should be given to relocating structures or facilities at risk of damage and to discouraging the public from frequenting the vicinity of potentially dangerous trees.

N.B. An attempt to arrest decay by excavating the rotten and infected wood may injure the defensive barriers around the infection and lead to even more extensive decay.

Preventing Decay

- Roots killed by other diseases or environmental conditions (waterlogging, compaction) and wounds to the underground parts (e.g. from trenching) are liable to infection from wood-rotting fungi, so such injuries should be avoided. However, some root-infecting, wood-rotting fungi spread vegetatively from neighbouring infected trees or stumps and can then penetrate live, uninjured roots. These can only be guarded against when they are known to be present and then only in certain circumstances (e.g. Honey fungus, p.145).
- Many decay fungi enter the tree when airborne spores germinate on stem or branch wood exposed by injuries to the bark. These should therefore be avoided as far as possible. Injuries which expose the end grain (e.g. pruning cuts) are particularly vulnerable.
- If pruning is unavoidable, it must not involve injury to the 'branch bark ridge' or the 'branch collar', if present. Broken and dead branches should be cut back with similar care to the nearest live branch junction or main stem (Lonsdale, 1999; Anon., 1989).
- Formative pruning when the tree is young and branches are small can obviate the later need for the removal of larger limbs.
- No available wound treatment is capable of protecting wounds against decay* for more than a year or so (Lonsdale, 1999).
- On present evidence, the increase in early callus growth brought about by certain wound paints is too short lived to warrant their general use.
- Theory and recent, limited short-term investigations indicate that wounds made in summer are less favourable for the development of early wound-colonizing, wood-rotting fungi than those made at other seasons, but whether season of wounding has a significant influence on decay development in the long term remains uncertain (Lonsdale, 1999).

*It may be beneficial to protect fresh pruning wounds on *Malus* and *Prunus* from the diseases Silver leaf (p.254) or Bacterial canker (p.72).

SECTION 6

REFERENCES AND MAIN LITERATURE SOURCES

† Main literature sources
* Arboriculture Research Notes are available from the Arboricultural Advisory and Information Officer, Alice Holt Lodge, Wrecclesham, FARNHAM, Surrey GU10 4LH, England. Research Information Notes are available from the Forestry Commission at the same address.

†ALFORD, D.V. (1991). *A colour atlas of pests of ornamental trees, shrubs and flowers*. Wolfe, London. 448 pages.

ANON [the current year]. *UK Pesticide Guide*. British Crop Protection Council and CAB International.

ANON [the current year]. *Pesticides* [the current year]. MAFF Reference Book 500. HMSO, London.

ANON (1989). *Recommendations for tree work*. British Standard 3998. British Standards Institution, London.

ANON (1990). *Amenity valuation of trees and woodlands*. Unnumbered leaflet. [The Helliwell System.] Arboricultural Association, Romsey, England.

†BARNES, H.G. (1951). *Gall midges of economic importance*. Gall midges of trees, vol. 5. Crosby Lockwood, London. 270 pages.

BARRATT, D.K.; SEABY, D.A.; GOURLAY, I.D. (1987). *A portable compression strength meter; a tool for the detection and quantification of decay in trees*. Arboricultural Journal 11, 313-322.

†BECKER, P. (1974). *Pests of ornamental plants*. MAFF Bulletin No. 97. HMSO, London.

†BEVAN, D. (1987). *Forest insects*. Forestry Commission Handbook No.1. HMSO, London.

BINNS, W.O. (1980). *Trees and water*. Arboricultural Leaflet No. 6. HMSO, London.

BURDEKIN, D.A. and RUSHFORTH, K.D. revised WEBBER, J.F. (1996). *Breeding elms resistant to Dutch elm disease*. Arboriculture Research Note 2/96/PATH.*

†BUTIN, H. (1995). *Tree diseases and disorders*. Oxford University Press

CARTER, C.I. (1971). *Conifer woolly aphids (Adelgidae) in Britain*. Forestry Commission Bulletin No.42. HMSO, London.

CARTER, C.I. (1972). *Winter temperatures and survival of the Green spruce aphid*. Forestry Commission Forest Record 84. HMSO, London.

CARTER, C.I. (1992). *Lime trees and aphids*. Arboriculture Research Note 104/92/ENT.*

CARTER, C.I. and MASLEN, N.R. (1982). *Conifer Lachnids*. Forestry Commission Bulletin 58. HMSO, London.

CARTER, C.I. & WINTER, T.G. (1998). *Christmas Tree Pests*. Forestry Commission Field Book 17. The Stationery Office, London.

†CHRYSTAL, R.N. (1937). *Insects of the British woodlands.* Warne, London. 370 pages.

†COOPER, J.I. (1979). *Virus diseases of trees and shrubs.* Institute of Terrestrial Ecology, Oxford. 74 pages.

†DARLINGTON, A. (1968). *The pocket encyclopaedia of plant galls in colour.* Blandford, London. 191 pages.

DAVIES, R.J. (1987). *Trees and weeds – weed control for successful tree establishment.* Forestry Commission Handbook No.2. HMSO, London.

DAVIES, R.J. and PEPPER, H.W. (1993). *Protecting trees from field voles.* Arboriculture Research Note 74/93/ARB.*

DOBSON, M.C. (1991a). *Diagnosis of de-icing salt damage to trees.* Arboriculture Research Note 96/91/PATH.*

DOBSON, M.C. (1991b). *Prevention and amelioration of de-icing salt damage to trees.* Arboriculture Research Note 100/91/PATH.*

DOBSON, M.C. (1991c). *Tolerance of trees and shrubs to de-icing salt.* Arboriculture Research Note 99/91/PATH.*

†DOBSON, M.C. (1991d). *De-icing salt damage to trees and shrubs.* Forestry Commission Bulletin 101. HMSO, London.

ERWIN, D.C. & RIBEIRO, O.K. (1996). *Phytophthora Diseases Worldwide.* APS Press.

FIELDING, N.J. (1992). *Rhizophagus grandis as a means of biological control against Dendroctonus micans in Britain.* Research Information Note 224. Forestry Commission, Edinburgh.*

GIBBS, J.N. (1992). *Crown damage to London plane.* Arboriculture Research Note 47/92/PATH.*

GIBBS, J.N. (1994). *De-icing salt damage to trees – the current position.* Arboriculture Research Note 119/94/PATH.*

GIBBS, J.N. (1999). *Dieback of pedunculate oak.* Forestry Commission Information Note 22. Forestry Commission, Edinburgh.

GIBBS, J.N., BRASIER, C.M. and WEBBER, J.F. (1994). *Dutch elm disease.* Research Information Note 252. Forestry Commission, Edinburgh.*

GIBBS, J.N. & GREIG, B.J.W. (1997). *Biotic and abiotic factors affecting the dying back of pedunculate oak, Quercus robur L.* Forestry, 70, 4, pp. 399-406.

GIBBS, J.N. & LONSDALE, D. (1998). *Phytophthora disease of alder.* Forestry Commission Information Note 6. Forestry Commission, Edinburgh.

GREGORY, S.C. & REDFERN, D.B. (1998). *Diseases and disorders of forest trees.* Forestry Commission Field Book 16. The Stationery Office, London.

GREIG, B.J.W. (1981). *Decay fungi in conifers.* Forestry Commission Leaflet No. 79. HMSO, London.

GREIG, B.J.W. (1992). *Occurrence of decline and dieback of oak in Great Britain.* Arboriculture Research Note 105/92/PATH.*

GREIG, B.J.W. (1994). *Dutch elm disease and English elm regeneration.* Arboriculture Research Note 13/94/PATH.*

GREIG, B.J.W., GREGORY, S.C. and STROUTS, R.G. (1991). Honey fungus. Forestry Commission Bulletin No. 100. HMSO, London.

†GRAM, E. and WEBER, A., edited by DENNIS, R.W.G. (1952). *Plant diseases.* Macdonald, London. 618 pages.

HARRIS, R.W. (1992). *Arboriculture.* Integrated management of landscape trees, shrubs and vines. Prentice-Hall, Englewood Cliffs, New Jersey.

HELLIWELL, D.R. (1990). *Acceptable levels of risk associated with trees.* Arboricultural Journal 14, 159-162.

HELLIWELL, D.R. (1993). *Water tables and trees.* Arboriculture Research Note 110/92/EXT.*

†HEPTING, G.H. (1971). *Diseases of forest and shade trees of the United States.* USDA Forest Service Agriculture Handbook No. 386. 658 pages.

HULL, S.K. and GIBBS, J.N. (1991). *Ash dieback - a survey of non-woodland trees.* Forestry Commission Bulletin No.93. HMSO, London.

†JEPPSON, L.R., KIEFER, H.H. and BAKER, E.W. (1975). *Mites injurious to economic plants.* University of California Press, Berkeley. 614 pages.

JOBLING, J. (1990). *Poplars for wood production and amenity.* Forestry Commission Bulletin No. 92. HMSO, London.

JUKES, M.R. (1984). *The Knopper gall.* Arboriculture Research Note 55/84/ENT.*

KING, C.J. and FIELDING, N.J. (1989). *Dendroctonus micans in Britain - its biology and control.* Forestry Commission Bulletin 85. HMSO, London.

LONSDALE, D. (1999). *Principles of Tree Hazard Assessment and Management.* Research for Amenity Trees No. 7. DETR/Forestry Commission. The Stationery Office, London.

LONSDALE, D. and WAINHOUSE, D.(1987). *Beech bark disease.* Forestry Commission Bulletin 69. HMSO, London.

†MANI, M.S. (1964). *Ecology of plant galls.* Junk, The Hague. 434 pages.

MATTHECK, C. & BRELOER, H. (1994). *The Body Language of Trees.* Research for Amenity Trees No. 4. Dept. of the Environment. HMSO, London.

†MEYER, J. (1987). *Plant galls and gall inducers.* Borntraeger, Berlin. 291 pages.

MITCHELL, A.F. (1973). *Replacement of elm in the countryside.* Forestry Commission Leaflet 57. HMSO, London.

†MITCHELL, A.F. (1974). *A field guide to the trees of Britain and Northern Europe.* Collins, London. 415 pages.

PATCH, D. and DENYER, A. (1992). *Blight to trees caused by vegetation control machinery.* Arboriculture Research Note 107/92/ARB.*

PATRICK, K.N. (1990). *Watermark disease of willows.* Arboriculture Research Note 87/90/EXT.*

†PEACE, T.R. (1962). *The pathology of trees and shrubs*. Oxford University Press. 723 pages.

PEPPER, H.W. (1988). *Protection against animals and management for game*. In: HIBBERD, B.G., (editor) Farm Woodland Practice, pages 46-57. Forestry Commission Handbook 3. HMSO, London.

†PHILLIPS, D.H. and BURDEKIN, D.A. (1982). *Diseases of forest and ornamental trees*. Macmillan, London. 435 pages. [A second, revised edition was published in 1992]

PHIPPS, R.L. (1985). *Collecting, preparing, crossdating and measuring tree increment cores*. U.S. Geological Survey, Water-Resources Investigations Report 85-4148.

PIEARCE, G.D. and GIBBS, J.N. (1981). *Verticillium wilt*. Arboricultural Leaflet No. 9. HMSO, London.

RISHBETH, J. (1982). *Bacterial wetwood*. Arboriculture Research Note 20/82/PATH.*

ROSE, D.R. (1983). *Cobweb fungus – Athelia*. Arboriculture Research Note 45/83/PATH.*

ROSE, D.R. (1989a). *Marssonina canker and leaf spot (anthracnose) of Weeping willow*. Arboriculture Research Note 78/89/PATH.*

ROSE, D.R. (1989b). *Scab and Black canker of willow*. Arboriculture Research Note 79/89/PATH.*

ROSE, D.R. (1990). *Lightning damage to trees in Britain*. Arboriculture Research Note 68/90/PATH.*

SCOTT, T.M. (1972). *The Pine shoot moth and related species*. Forestry Commission Forest Record 83. HMSO, London.

SHIGO, A.L. (1986). *A new tree biology*. Shigo and Trees Associates, Durham, New Hampshire.

†SINCLAIR, W.A.; LYON, H.H.; JOHNSON, W.T. (1987). *Diseases of trees and shrubs*. Cornell University Press, Ithaca. 574 pages.

†SMITH, I.M.; DUNEZ, J.; LELLIOTT, R.A.; PHILLIPS, D.H.; ARCHER, S.A. (editors) (1988). *European handbook of plant diseases*. Blackwell Scientific Publications, Oxford. 583 pages.

STROUTS, R.G. (1981). *Phytophthora diseases of trees and shrubs*. Arboricultural Leaflet No. 8. HMSO, London.

STROUTS, R.G. (1990). *Coryneum canker of Monterey cypress and related trees*. Arboriculture Research Note 39/90/PATH.*

STROUTS, R.G. (1991a). *Anthracnose of London plane*. Arboriculture Research Note 46/91/PATH.*

STROUTS, R.G. (1991b). *Dieback of the Flowering cherry, Prunus 'Kanzan'*. Arboriculture Research Note 94/91/PATH.*

STROUTS, R.G. and PATCH, D. (1994). *Fireblight on ornamental trees and shrubs*. Arboriculture Research Note 118/94/PATH.*

†STUBBS, F.B. (editor) (1986). *Provisional keys to British plant galls.* British Plant Gall Society, Leicester. 96 pages.

TATTAR, T.A. (1978). *Diseases of Shade Trees.* Academic Press, New York.

WAINHOUSE, D. (1994). *The Horse chestnut scale: a pest of town trees.* Arboriculture Research Note 122/94/ENT*.

†WILSON, G.F. (1960). *Horticultural pests.* Detection and control. Crosby Lockwood, London. 240 pages.

WINTER, T.G. (1983). *A Catalogue of Phytophagous Insects and Mites on Trees in Great Britain.* Forestry Commission Booklet 53. Forestry Commission, Edinburgh.

WINTER, T.G. (1989). *Cypress and juniper aphids.* Arboriculture Research Note 80/89/ENT.*

WINTER, T.G. (1991). *Pine shoot beetles and ball-rooted semi-mature pines.* Arboriculture Research Note 101/91/ENT.*

†WORMALD, H. (1955). *Diseases of fruits and hops.* Crosby Lockwood, London. 325 pages.

YOUNG, C.W.T. (1978). *Sooty bark disease of sycamore.* Arboricultural Leaflet No. 3. HMSO, London.

APPENDIX 1

HINTS ON COLLECTING AND DISPATCHING SPECIMENS FOR LABORATORY EXAMINATION

Before submitting any specimens it is as well to discuss the matter with the laboratory on the telephone.

All specimens: Send enough specimens to show all stages of the disorder, from apparently healthy to severely damaged tissues. Pack and dispatch soon after collection together with a full account of the problem. Send early in the week by the quickest method as delay may render specimens useless. Allow any superficial moisture on the specimens to dry before packing and add no moisture to packages (but see 'insects' below). If in any doubt over the identity of the host tree, include an appropriate specimen of it.

Small trees: If trees are small enough to send whole, dig them up (the damage caused by pulling them up can make diagnosis more difficult), gently shake off loose soil (retaining any required as a specimen) and pack the trees in polythene bags. Cut the trees into pieces if this aids packing but label the pieces so that each tree can be reconstructed.

Root, stem and **branch specimens:** Include the junction between any dead and live bark.

Soil specimens to be checked for Phytophthora: Take small samples of soil from near roots and stems at scattered points at various depths around affected trees to make, in total, at least a cupful. Pack in a sealed polythene bag.

Foliage, woody material and **hard, dry fungal fruit bodies:** Pack in sealed polythene bags. If leaf-fall is the problem, send some fallen leaves together with some twigs with affected leaves still attached. Samples of decayed wood should be cut to include the zone between decayed and sound wood, where possible.

Fleshy or hard, moist fungal fruit bodies: Send in dry packing material in a stout cardboard box. Do not use a polythene bag or tin as such specimens putrefy quickly if kept damp. If a delay is unavoidable before dispatch, keep specimens in an open container in a refrigerator or in a dry room. A spore print from toadstools can be useful: place an open cap, gills down, on a piece of white and another on a piece of black paper for a few hours.

Insects etc.: Send insects and the like in rigid, escape-proof containers with a little soft, dry tissue inside (slightly moistened if the insect is alive) to secure the specimens. Include a little foodstuff as appropriate. Secure the lids with sticky tape.

APPENDIX 2

SOURCES OF HELP IN THE DIAGNOSIS OF DAMAGE TO AMENITY TREES AND ON RELATED MATTERS

ADVISORY SERVICES WITH LABORATORY AND OTHER SPECIALIST FACILITIES

a. **The Arboricultural Advisory and Information Service** (AAIS),
   Alice Holt Lodge, Wrecclesham, FARNHAM, Surrey GU10 4LH.
   Tree Helpline Tel: 09065-161147. Fax: 01420-22000.

b. **The Forestry Commission's Research Stations**:
   (a) Alice Holt Lodge, Wrecclesham, FARNHAM, Surrey GU10 4LH.
       Tel: 01420-22255. Fax: 01420-23653.
   (b) Northern Research Station, ROSLIN, Midlothian EH25 9SY.
       Tel: 0131-445 2176. Fax: 0131-445 5124.

c. **The Royal Horticultural Society**, (for RHS Members only)
   RHS Garden, Wisley, WOKING, Surrey GU23 6QB.
   Tel: 01483-224234. Fax: 01483-211750.

SOURCES OF INFORMATION ON PLANT HEALTH LEGISLATION

d. Central Science Laboratory,
   Ministry of Agriculture, Fisheries and Food,
   Sand Hutton, YORK YO41 1LZ.
   Tel: 01904-462000

e. Scottish Executive Rural Affairs Department,
   Pentland House, 47 Robb's Loan, EDINBURGH EH14 1TW.
   Tel: 0131-556 8400.

f. Agriculture Policy Division, National Assembly for Wales
   Crown Buildings, Cathays Park, CARDIFF CF10 3NQ.
   Tel: 01222-825111.

g. Plant Health Service, Forestry Commission,
   231 Corstorphine Road, EDINBURGH EH12 7AT.
   Tel: 0131-334 0303.

GLOSSARY

Plurals not ending in –s are given in round brackets.

acaricide	Pesticide used to control mites.
adelgid	Sap-sucking insect in the family Adelgidae (Order Hemiptera). All species feed on conifers and most produce waxy wool. See p.253 for differences from aphids and phylloxerids.
aeciospore	One of the several different kinds of spores produced asexually by a rust fungus.
aerial	Above ground.
agromyzid	Two-winged fly in the family Agromyzidae (Order Diptera). The larvae mine in plant tissue.
alate	Winged form of an insect; a term used most frequently with regard to aphids. Cf. apterous.
allelopathy	see p. 6
anamorph	The asexual fruiting form of a fungus. See teleomorph.
anthracnose	A disease characterized by small, limited, black lesions.
apterous	Wingless form of an insect. Cf. alate.
arthropod	A creature in the Arthropoda ('with jointed legs'), which includes the insects and mites.
ascospore	Sexually produced spore of a fungus in the taxonomic Division Ascomycotina, which includes most of the micro-fungi which cause tree diseases.
axial	In the direction of the axis or pith; longitudinal.
basidiospore	Sexually produced spore of a fungus in the taxonomic Division Basidiomycotina, which includes the gill fungi, the rusts and most of the bracket fungi.
blight	A loose term describing the extensive and rapid death and collapse of soft tissues.
bracket fungus	A fungus whose fruit bodies resemble brackets, shelves or hoofs, e.g. Piptoporus, Coriolus, Fomes. (see **Figs. 348, 336, 340**).
calcicole	Of plants which grow best on calcareous (alkaline) soils. Cf. calcifuge.
calcifuge	Of plants which grow poorly on calcareous (alkaline) soils. Cf. calcicole.
canker	A clearly defined patch of dead and sunken or malformed bark. *Annual* c.: one which does not enlarge after its first year. *Erumpent* c.: one consisting of an irregular mass of malformed, often thickened bark protruding from the normal bark, as in Bacterial canker of ash, (**Fig.28**). *Perennial* c.: one which enlarges in the years following its year of initiation. *Sunken* c.: one where subsequent callus growth has elevated the surrounding bark, leaving the necrosis as a depression. *Target* c.: a sunken perennial canker consisting of concentric ridges of killed callus tissue (**Fig.259**).

canker-rot
: A disease in which the causal fungus gives rise to both a bark canker and to decay in the underlying wood (see e.g. **Figs 382-384**).

cecidomyid
: Two-winged fly in the family Cecidomyiidae (Order Diptera); many species in this group are gall midges.

chlorosis
: Abnormal yellow or yellow-green coloration of normally green foliage.

coccid
: Soft or waxy scale insect in the Family Coccidae (Order Hemiptera).

conidium (conidia)
: An asexually produced spore. Cf. basidiospore, ascospore.

crawler
: Mobile first stage larva (nymph) of a scale insect (the later stages and adults of many scales are sedentary).

cynipid
: Gall wasp in the Family Cynipidae (Order Hymenoptera); most British species cause galls on Quercus.

diapause
: Period of suspended development whereby an insect can remain in the same stage for longer than normal.

dieback
: Often loosely used to mean 'death'. Here used to mean the progressive death of a tree or branch from its extremities towards the roots.

distal
: The part (of a limb, root, leaf, etc.) furthest from the point of attachment. Cf. proximal.

endophytic fungus
: A fungus which can survive inactive and harmless inside a plant. Such fungi may cause disease if conditions suitable for their active growth arise.

eriophyid
: Mites in the Superfamily Eriophyoidea (Subclass Acari). Includes rust, blister and gall mites. The last are mainly in the family Eriophyidae, have a sausage-shaped body about 0.1–0.2mm long and four legs at the anterior (front) end.

frass
: Bore dust, excrement and other debris left by bark beetles and other feeding insects.

fruit body
: A general term for any kind of fungal, spore-bearing structure (e.g. **Figs. 316-352, 99, 209, 270**)

gall
: Abnormal plant growth. See p.129.

gill fungus
: A fungus whose fruit bodies are toadstools like the edible mushroom, with gills beneath the caps. See e.g. **Fig.183**.

girdling
: Encircling the organ in a broad or narrow band. A girdled stem, branch or shoot is encircled by a band of dead, dying, missing or constricted bark (**Figs. 11, 237**). The distal part then usually dies.

honeydew
: Sugary solution excreted by sap-sucking insects. See p.140.

host (tree)
: The tree on or in which the parasite lives.

hypha (-e)
: One of the extremely fine threads of which a fungus consists.

imperfect
: (fruit bodies, spores). See teleomorph.

incipient decay
: Wood infected (and often stained) by a wood-rotting fungus but not yet evidently decayed.

infection court
: the point on the plant where infection takes place

larva (-e)	Immature stage of an insect – the types commonly known as caterpillars or grubs.
leafhopper	Member of the insect family Cicadellidae (Order Hemiptera); the tree-feeding species belong to several Subfamilies but mostly to the Typhlocybinae.
lesion	A localized area of diseased or disordered tissue.
longhorn beetle	Member of the Family Cerambycidae (Order Coleoptera); the adults have long antennae; the larvae of most species tunnel in wood.
mesophyll cell	Chlorophyll-containing cell of the leaf, between the upper and lower leaf surfaces.
mycelium (mycelia)	A mass of hyphae (often readily visible to the naked eye).
necrosis	Death of plant tissues, usually characterized by a change in colour to brown or black. (adjective necrotic)
nothovar	(in plant names). A variety of a hybrid.
nymph	Larva or immature stage of an insect such as a leafhopper, aphid or psyllid which changes into an adult by developing reproductive organs and usually wings, rather than by changing into a pupa from which the adult emerges, as do beetles, butterflies and moths.
occlusion	The overgrowth of a wound with (callus) tissue produced subsequently. (verb: occlude)
oedema	The growth of small corky lumps on the underside of leaves due to an excess of water in the plant. See Waterlogging p.266.
oospore	A sexually produced resting spore of Phytophthora species and related fungi.
ovipara (-e)	Sexually reproducing female of an aphid; the form that lays the eggs which overwinter.
ovipositor	In insects, a tube-like structure at the tip of the female abdomen through which the eggs are laid.
parenchyma	Soft, living plant tissue consisting of simple thin-walled cells with intervening air spaces.
parthenogenesis	Type of reproduction in which unfertilized females produce viable eggs or young. Common among aphids and closely related groups.
pathogen	An organism which causes disease (adjective: pathogenic).
pathovar	A variety of a pathogenic species. Often used where the main feature distinguishing it from the species is its host range. (Abbreviated to pv.)
perennate	(of a canker, lesion or disease). To continue to enlarge or develop year after year.
perfect	(fruit bodies, spores). See teleomorph.
pesticide	A material for killing a pest or pathogen.
phylloxerid	Sap-sucking insect in the family Phylloxeridae (Order Hemiptera). See p.253 for difference from adelgids and aphids.
phytotoxic	Poisonous to plants.

proximal   The part (of the limb, root, leaf etc.) nearest the point of attachment. Cf. distal.

psyllid   Member of the Family Psyllidae (Order Hemiptera). Many species cause galls or other leaf deformities. See pp.84 and 130.

pupa (-e)   Immobile, non-feeding stage of an insect between larva and adult when many internal changes take place (metamorphosis); chrysalis.

puparium   Hard case formed by many flies (Diptera) from the last larval skin and within which the pupa is formed.

pycnidium (pycnidia)   An enclosed fruit body, containing conidia, of some microfungi.

rostrum   Visible mouthparts of a sucking insect such as an aphid, leafhopper or scale insect.

rust fungus   A pathogenic fungus in the Order Uredinales, so called from the conspicuous rust-coloured or yellow spore masses it produces. See e.g. **Fig.353**.

rust mite   See eriophyid.

saprophyte   An organism which subsists on dead plant material. Some prefer the term 'saprobe' when referring to fungi and bacteria as these are, strictly speaking, not 'phyta' (plants).

scab   (i) A plant disease characterized by roughened, raised, scab-like patches on infected tissues; (ii) a disease not characterized by scabs but caused by fungi closely related to Venturia inaequalis, the cause of Apple scab. See p.225.

siphuncular cone   One of a pair of conical structures (cornicles) situated on the dorsal (upper) surface of the abdomen (5th or 6th segment) of an aphid.

sooty mould   A black Saprophytic fungus growing on honeydew. See p.140.

spider mite   Small, active, silk-spinning mite (Order Acariformes, Family Tetranychidae) with either eight legs (adult) or six legs (immature stages). It feeds on the leaf surface by sucking up the contents from cells it has pierced with its mouthparts. See pp.103 and 167.

spore tendrils   Thread-like masses of spores extruded from the fruit bodies of some micro fungi. See e.g. **Fig.112**.

stag-headed   Describes the silhouette of a large tree whose crown has died back so that the ends of the dead branches protrude like spikes or antlers from the reduced live foliated crown (see **Fig.7**).

substrate   Plant tissue on or in which another organism is growing.

sucker   See psyllid.

teleomorph   The sexual fruiting form of a fungus. Many fungi produce spores both by fusion of male and female sex cells (the perfect state) and vegetatively (the imperfect state or states). As each fruiting form or state receives a name according to its morphological and developmental characteristics, one fungus often has two valid names and sometimes more. See anamorph.

teliospore	Spore of a rust fungus which (usually) carries the fungus through the winter and gives rise to infective basidiospores in the spring.
therapy	Treatment of an existing disease or disorder.
tortrix	Small moth in the Family Tortricidae (Order Lepidoptera) (adjective tortricid).
urediniospore	An asexual spore of a rust fungus which germinates usually without an overwintering or resting period. These are therefore sometimes called summer spores.
vector	An organism that carries a pathogen to a plant.
viviparous	Giving birth to live young.
volva	A membranous cup out of which the stems of some toadstools grow (e.g. Amanita species).
weevil	Beetle with characteristic elbowed antennae usually borne on a snout (Family Curculionidae, Order Coleoptera).
yeast-like	Consisting of an amorphous spore mass which has arisen as the result of the repeated division of single-celled spores.
zoospore	An asexually produced, infective, motile spore of Phytophthora species and related fungi.

SECTION 6

INDEX

NB
1   TREE NAMES: Only English names are listed alphabetically here; these are cross referenced to the Latin names in Section 2.
2   Headwords in lower case letters are explained in the Glossary in Section 6.
3   'see' denotes synonymy.
4   Page numbers in bold type are main references.

SECTION **6**

For Latin tree names see Section 2

For Latin tree names see Section 2

SECTION **6**

SECTION 6

For Latin tree names see Section 2

SECTION 6

For Latin tree names see Section 2

For Latin tree names see Section 2

SECTION 6

For Latin tree names see Section 2

SECTION 6

SECTION **6**

SECTION **6**

For Latin tree names see Section 2

For Latin tree names see Section 2

SECTION 6

For Latin tree names see Section 2

For Latin tree names see Section 2

SECTION 6

For Latin tree names see Section 2

NOTES

Printed in the United Kingdom for The Stationery Office
N167143   C10   06/04

A DIAGNOSTIC PROCEDURE

- Start at box 1 and follow the appropriate arrows.
- To avoid a wrong or incomplete diagnosis, do not omit any steps.
- Remember that several different problems may be evident at once.
- For more information, see p.16

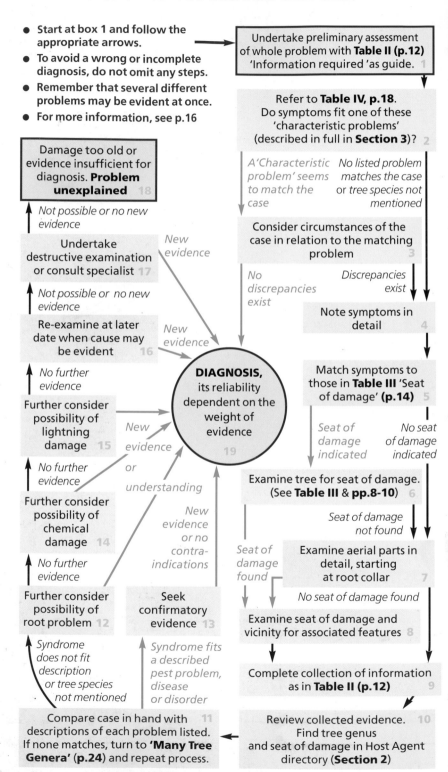

Undertake preliminary assessment of whole problem with **Table II (p.12)** 'Information required 'as guide. **1**

Refer to **Table IV, p.18**. Do symptoms fit one of these 'characteristic problems' (described in full in **Section 3**)? **2**

A 'Characteristic problem' seems to match the case *No listed problem matches the case or tree species not mentioned*

Consider circumstances of the case in relation to the matching problem **3**

No discrepancies exist *Discrepancies exist*

Note symptoms in detail **4**

Match symptoms to those in **Table III** 'Seat of damage' **(p.14)** **5**

Seat of damage indicated *No seat of damage indicated*

Examine tree for seat of damage. (See **Table III** & **pp.8-10**) **6**

Seat of damage not found

Seat of damage found Examine aerial parts in detail, starting at root collar **7**

No seat of damage found

Examine seat of damage and vicinity for associated features **8**

Complete collection of information as in **Table II (p.12)** **9**

Review collected evidence. **10** Find tree genus and seat of damage in Host Agent directory **(Section 2)**

Compare case in hand with **11** descriptions of each problem listed. If none matches, turn to **'Many Tree Genera' (p.24)** and repeat process.

Syndrome does not fit description or tree species not mentioned *Syndrome fits a described pest problem, disease or disorder*

Seek confirmatory evidence **13**

Further consider possibility of root problem **12**

No further evidence

Further consider possibility of chemical damage **14**

No further evidence

Further consider possibility of lightning damage **15**

No further evidence

Re-examine at later date when cause may be evident **16**

Not possible or no new evidence

Undertake destructive examination or consult specialist **17**

Not possible or no new evidence

Damage too old or evidence insufficient for diagnosis. **Problem unexplained 18**

New evidence

New evidence

DIAGNOSIS, its reliability dependent on the weight of evidence **19**

New evidence or understanding

New evidence or no contra-indications